Lecture Notes in Computer Science 13597

Founding Editors

Gerhard Goos

Juris Hartmanis

Editorial Board Members

The series Lecture Notes in Computer Science (LNCS), including its subseries Lecture Notes in Artificial Intelligence (LNAI) and Lecture Notes in Bioinformatics (LNBI), has established itself as a medium for the publication of new developments in computer science and information technology research, teaching, and education.

LNCS enjoys close cooperation with the computer science R & D community, the series counts many renowned academics among its volume editors and paper authors, and collaborates with prestigious societies. Its mission is to serve this international community by providing an invaluable service, mainly focused on the publication of conference and workshop proceedings and postproceedings. LNCS commenced publication in 1973.

Bin Sheng · Marc Aubreville

Editors

Mitosis Domain Generalization and Diabetic Retinopathy Analysis

MICCAI Challenges MIDOG 2022 and DRAC 2022
Held in Conjunction with MICCAI 2022
Singapore, September 18–22, 2022
Proceedings

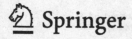 Springer

Editors
Bin Sheng
Shanghai Jiao Tong University
Shanghai, China

Marc Aubreville 🆔
Technische Hochschule Ingolstadt
Ingolstadt, Germany

ISSN 0302-9743   ISSN 1611-3349 (electronic)
Lecture Notes in Computer Science
ISBN 978-3-031-33657-7   ISBN 978-3-031-33658-4 (eBook)
https://doi.org/10.1007/978-3-031-33658-4

This Springer imprint is published by the registered company Springer Nature Switzerland AG
The registered company address is: Gewerbestrasse 11, 6330 Cham, Switzerland

Preface

This volume comprises in total 25 scientific papers (20 long papers and five short papers) that have all undergone peer review from the following two biomedical image analysis challenges at MICCAI 2022: the Diabetic Retinopathy Analysis Challenge (DRAC 2022) and the Mitosis Domain Generalization Challenge (MIDOG 2022).

Our challenges share the need for developing and fairly evaluating algorithms that increase accuracy, reproducibility, and efficiency of automated image analysis in clinically relevant applications.

The DRAC 2022 challenge aimed at providing three clinically relevant tasks for diabetic retinopathy (DR) analysis with Ultra-Wide Optical Coherence Tomography Angiography (UW-OCTA) Images. Segmentation of DR lesions, image quality assessment, and DR grading were addressed by a variety of state-of-the-art algorithms. The challenge was organized using the grand-challenge.org platform, where the challenge submission and automatic evaluation were performed.

The MIDOG 2022 challenge sought to identify domain-agnostic approaches for the detection of mitosis in histopathological images, a critical component of breast cancer diagnosis and grading typically conducted by pathologists. Compared to the preceding 2021 MIDOG event, the primary innovation in this year's challenge was the incorporation of a broader range of domain variability, including variations caused by a change in laboratories, species, and tumor types. Participants submitted their algorithms in a Docker container format via the grand-challenge.org platform.

Bin Sheng
DRAC 2022 General Chair

Marc Aubreville
MIDOG 2022 General Chair

Contents

MIDOG

DRAC

DRAC 2022 Preface

The computer-aided diagnostic system is of great importance in clinical disease screening and analysis. Over the past few years, deep learning has shown great success in medical image analysis, contributing to the improvement of diagnostic efficiency. In particular, computer-assisted automatic analysis of diabetic retinopathy (DR) is essential in reducing the risks of vision loss and blindness. The tasks where automatic algorithms are especially helpful within the diagnostic process of DR include segmentation of DR lesions, image quality assessment and DR grading. With the help of deep learning methods, the workload of ophthalmologists can be reduced, and routine screening is possible, especially in rural areas and areas where medical resources are scarce. At present, there are already some challenges that provide benchmarks for the evaluation of fundus photography for DR analysis. However, as an important imaging modality in DR diagnosis system that can non-invasively detect the change of neovascularization, ultra-wide optical coherence tomography angiography (UW-OCTA) has attracted a little attention from challenge organizers. There is a lack of publicly available UW-OCTA datasets for fair performance evaluation of DR diseases. To address this limitation, we organized the Diabetic Retinopathy Analysis Challenge (DRAC 2022), which was a part of the 25th International Conference on Medical Image Computing and Computer Assisted Intervention (MICCAI 2022). This challenge allowed for the first time an analysis of method performance for diabetic retinopathy with the UW-OCTA imaging modality. To date, the challenge website remains open for post-challenge registrations and submissions, aiming to provide a sustainable benchmark and evaluation platform for more algorithms.

The challenge comprised three clinically relevant sub-tasks: segmentation of DR lesions, image quality assessment and DR grading. In the task of DR lesion segmentation, there are 109 images for training and 65 images for testing. In the task of image quality assessment, there are 665 images for training and 438 images for testing. In the task of DR grading, there are 611 images for training and 386 images for testing. The challenge submission and automatic evaluation were performed on the grand-challenge.org platform (drac22.grand-challenge.org). Participants can have access to the dataset after they register on the Grand Challenge website and sign the challenge rule agreement consent form. In addition, participants can browse the challenge rules and news, examine challenge timelines and find their rankings on the challenge website. Dice Similarity Coefficient (DSC) is used for the algorithm evaluation and ranking in the segmentation task. Quadratic weighted kappa is used for the algorithm evaluation and ranking in the two classification tasks. In addition, each team was required to submit a paper reporting their challenge results and methods from the perspective of data preprocessing, data augmentation, model architecture, model optimization and post-processing. When participating in multiple tasks, each team could either submit several papers or a single paper reporting all methods. Finally, we received eighteen papers from the participating teams, where 11, 12 and 13 different methods were reported for the tasks of

DR lesion segmentation, image quality assessment and DR grading respectively. These teams were eligible for the final official ranking, and some of the teams were invited to give a presentation at a half-day satellite event of MICCAI 2022 held on September 18, 2022.

This proceeding contains eighteen selected papers that cover a wide variety of state-of-the-art deep learning methods for different challenge tasks. All papers underwent a single-blind peer review with a minimum of three reviewers per paper, checking for novelty and quality of the work. We would like to thank all the DRAC 2022 participants, committee members and reviewers for their efforts and contributions to the success of this challenge.

December 2022

Bin Sheng
Huating Li
Hao Chen

DRAC 2022 Organization

General Chairs

Bin Sheng   Shanghai Jiao Tong University, China
Huating Li   Shanghai Sixth People's Hospital, China
Hao Chen   Hong Kong University of Science and
     Technology, China

Program Committee

Weiping Jia   Shanghai Sixth People's Hospital, China
Yiyu Cai   Nanyang Technological University,
     Singapore
Qiang Wu   Shanghai Sixth People's Hospital, China
Xiangning Wang   Shanghai Sixth People's Hospital, China
Bo Qian   Shanghai Jiao Tong University, China
Ruhan Liu   Shanghai Jiao Tong University, China
Ling Dai   Shanghai Jiao Tong University, China
Haoxuan Che   Hong Kong University of Science and
     Technology, China
Ping Zhang   Ohio State University, USA

Additional Reviewers

Jin Huang
Mengsi Guo
Wenqing Zhao

nnU-Net Pre- and Postprocessing Strategies for UW-OCTA Segmentation Tasks in Diabetic Retinopathy Analysis

Felix Krause[1(✉)], Dominik Heindl[1], Hana Jebril[2], Markus Karner[1], and Markus Unterdechler[1]

[1] Johannes Kepler University Linz, Linz, Austria
krause@ml.jku.at, dominik@heindl.one, m.unterdechler@gmx.at
[2] Department of Ophthalmology, Medical University of Vienna, Vienna, Austria
hana.jebril@meduniwien.ac.at

Abstract. Ultra-wide (UW) optical coherence tomography angiography (OCTA) imaging provides new opportunities for diagnosing medical diseases. To further support doctors in the recognition of diseases, automated image analysis pipelines would be helpful. Therefore the MICCAI DRAC 2022 challenge was carried out, which provided a standardized UW (swept-source) OCTA data set for testing the effectiveness of various algorithms on a diabetic retinopathy (DR) dataset. Our team tried to train well-performing segmentation models for UW-OCTA analysis and was finally ranked under the three top-performing teams for segmenting DR lesions. This paper, therefore, summarizes our proposed strategy for this task and further describes our approach for image quality assessment and DR Grading.

Keywords: UW-OCTA · segmentation · diabetic retinopathy · MICCAI DRAC 2022 challenge

1 Introduction

Artificial intelligence (AI) achieved great success in medical image analysis. Machine learning algorithms have been developed that can detect diseases from medical images at an expert level [1,2,4]. However, medical imaging and detectable abnormalities come in different forms, and each have their own challenges and opportunities. It is all the more important that the attention is drawn to a borader community with different approaches and ideas. A possible way to push state-of-the-art is to organize international challenges, where participants can compete against each other and compare their methods on a public leaderboard.

Diabetic retinopathy (DR) is an eye disease that affects approximately 78% of people with a history of diabetes of more than 15 years and is a leading cause of blindness [12]. The disease can be classified into mild non-proliferative

B. Sheng and M. Aubreville (Eds.): MIDOG 2022/DRAC 2022, LNCS 13597, pp. 5–15, 2023.
https://doi.org/10.1007/978-3-031-33658-4_1

diabetic retinopathy (NPDR) and the more advanced stage proliferative diabetic retinopathy (PDR). Abnormal retinal lesions like intraretinal microvascular anomalies (IrMAs), neovascularizations (NV), and areas of non-perfusion (NPAs) are used to determine the progress of DR. IrMAs are shunt vessels, one of the characteristics of end-stage NPDR [3], and appears as abnormal branching or dilation of existing blood vessels. NV appears as abnormal blood vessel growth [14] and determines whether the DR is proliferative (with NV) or not (without NV).

The Diabetic Retinopathy Analysis Challenge 2022 (DRAC22) [10] addresses the challenges of a fully automated analysis of eye scans. The eye scans are made using ultra-wide optical coherence tomography angiography imaging (UW-OCTA), which has already demonstrated significant advantages over typical OCT in DR detection [15]. Fully automated methods must be found to assess the quality of the eye scans, to segment retinal lesions (IrMAs, NV and NPAs) and to classify the grade of a possible diabetic retinopathy (DR).

2   Details on the MICCAI DRAC 2022 Challenge

The Diabetic Retinopathy Analysis Challenge 2022 (DRAC22) is associated with the International Conference on Medical Image Computing and Computer-Assisted Intervention (MICCAI) [10]. It is about analyzing eye scans with ultra-wide optical coherence tomography angiography (UW-OCTA) for diabetic retinopathy (DR).

The eye scans were provided as grayscale PNG files with a resolution of 1024×1024 pixels.

The challenge consists of 3 tasks, which were evaluated and ranked independently of each other. The **Image Quality Assessment** is a classification task. Here, the model has to predict the correct quality class of a given image: poor, good, or excellent. The **Diabetic Retinopathy Grading** is also a classification task, where the progression of DR has to be predicted. It is described in 3 classes: an undamaged normal eye, mild non-proliferative diabetic retinopathy (NPDR), or the more advanced stage proliferative diabetic retinopathy (PDR).

Table 1. Size of training and validation data per task.

	Task	Training data size	Validation data size
Task 1	Segmentation of DR Lesions	109	65
Task 2	Image Quality Assessment	665	438
Task 3	DR Grading	611	386

The **Segmentation of Diabetic Retinopathy Lesions** task consists of 3 different DR lesions segmentation sub-tasks, where the prediction results were averaged for the final ranking. In these sub-tasks, IrMAs, NPAs, and NV have to be found and segmented.

In total, we got a train dataset containing 665 images. However, the ground truths were not given for all of these images in the tasks. In Table 1 an overview of the available amount of labeled training and validation images is shown. All 665 images are labeled for the Image Quality Assessment and 611 are also labeled for the DR Grading. Much less training data was available for the segmentation tasks.

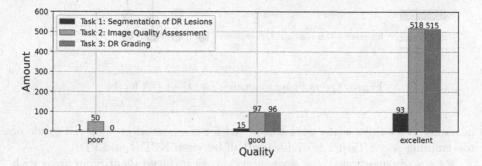

Fig. 1. Amount of training data per quality.

Figure 1 shows that 50 images were of the lowest quality, and only one image with bad quality was used in another task. Generally, we can see that most of the available training data for each task are of the best quality.

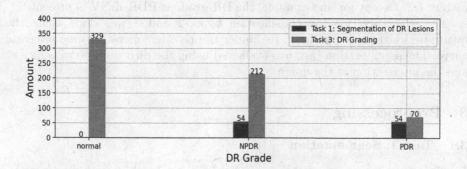

Fig. 2. Amount of training data per DR grade.

A closer look at the DR grade shows that more than half of the training set was labeled as normal, and only a small proportion of PDR is present (see Fig. 2). In combination with the segmentation task, we can see that none of the

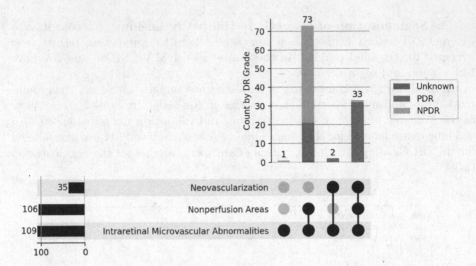

Fig. 3. Overlapping anomalies and their DR levels.

normally graded eye scans were considered for segmentation. The DR grade of the training set of Task 1 is divided in half between NPDR and PDR.

As already mentioned, the segmentation task involved identifying areas with anomalies in the given training images. However, not all anomalies were present in every image. IrMAs can be found in every image, followed by NPAs, only missing in a small set. NV is only visible in 35 of the 109 examples.

Figure 3 shows the occurrence of anomalies and their overlap along with the corresponding DR grade. A closer look shows that a combination of IrMAs and NPAs represents the majority. The DR grade also appears to be distributed according to the underlying evidence, with a PDR - NPDR ratio of approximately 1:3. Except for one example, the DR grade is PDR if NV is present.

The evaluation for the classification tasks 2 and 3 was done using the quadratic weighted kappa and area under the receiver operating characteristic curve. The segmentation task was evaluated using the dice similarity coefficient and the intersection of the union.

3 Preprocessing

3.1 Task 1: Segmentation

For the segmentation, we used the nnU-Net framework, which includes a preprocessing strategy that automatically adapts itself to the data at hand. nnU-Net therefore offers a holistic configuration with the need of expert knowledge. Even though the framework mostly focuses on the application on 3D data it can still be used successfully for 2D data. The automatic configuration of the nnU-Net is done in three optimization steps based on fixed, rule-based and empirical parameters. While the first step contains parameters for the learning process and the U-Net architecture itself, the rule-bases optimization applies heuristic rules that connect dataset properties and pipeline design choices. Those dataset specific fingerprints consist of image modality, intensity distribution, distribution of spacings and shape of the data. Those parameters make up the basic information about the dataset which then get processed further, until a decision about batch size, patch size, and final network topology is made. This iterative process tries to maximize the available GPU memory to allow for maximum batch and patch size. While larger batch sizes improve gradient estimates and therefore impact performance, the nnU-Net framework prioritizes the patch size to increase the contextual information available to the network [6,9]. Albeit the applied strategy is traceable, it is too complex to be set out in detail in this article. For further explanation, we refer to Isensee et al. [6].

3.2 Task 2 and Task 3: Image Quality Assessment and Diabetic Retinopathy Grading

For Task 2 and Task 3, we used similar approaches, which resulted in different models for a combined ensemble approach. While some models were standard pre-trained versions of ConvNext [7] or EfficientNet V2 [11], which did not need any preprocessing, we also used an adapted version of deep multiple instance learning (MIL) for high-resolution images. For more details on this method, refer to Ilse et al. [5]. In short, this approach slices the original image into smaller patches which can then be fed into a neural network, where the output of the network is saved as a low-dimensional embedding of the original slice. The embeddings of all slices of the original image are then combined into a bag and labeled with a single label. These bags are then fed into a small attention network as described in Ilse et al. [5].

4 Proposed Image Segmentation Strategy

4.1 Baseline

For this task, we applied the 2D nnU-Net without modification to the published data to establish a baseline for reference [6]. Due to the difference in available samples per class, we train a separate model on each class, which allows for class-specific experimentation and fine-tuning. Except for the changes mentioned in Sect. 4.2, all models were trained in the same environment. For the training, a combination of Dice and Cross-Entropy Loss was used (Dice + CE Loss), whereas for evaluation, the Dice-score provided by the nnU-Net framework was applied. Furthermore, we chose the SGD optimizer with a momentum of 0.99 as well as a polynomial learning rate scheduler. The implementation can be found on our GitHub page[1]. We used the standard 5-fold cross-validation from the nnU-Net framework, resulting in the training of five different models. Furthermore, we trained for 200 epochs, because at this stage every model had clear signs of

---

[1] https://github.com/flixmk/DRAC22-JKU.

overfitting. It is important to mention here that in every run, no matter the modifications, the five models always performed very differently, with roughly half of them being successful while others tended to overfit from a very early stage onwards. The effects of this are demonstrated in Fig. 4. Over the course of the challenge, this was the most significant problem our team faced, because it did not allow for continuity and reproducibility. Consequently, many successes using different modifications could not be traced back to individual adjustments. Furthermore, it raises doubt about to what extent the applied changes can be trusted. Repeatedly training on the same train/validation data split produces very similar results, which leads us to believe that the results are, at least in a small fraction, dependent on what samples were used for training or validation respectively. Unfortunately, in the end, we were not able to prove this hypothesis by manual inspection. However, we frequently found reoccurring samples in the validation and training set in poorly performing folds. Due to a lock of trials we could not attribute any statistical significance.

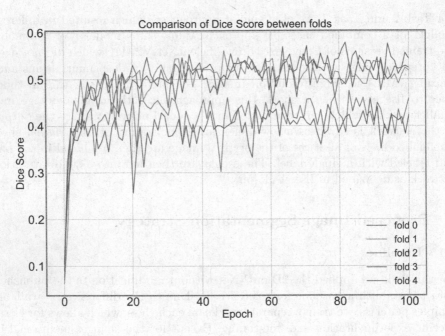

Fig. 4. Comparison of the five folds trained during a run using the Dice score as evaluation. It can be seen that two folds perform significantly worse than the others hinting at the just explained problem.

4.2   nn-UNet Modifications

Data Augmentation. To avoid overfitting the training data, we applied a thorough data augmentation (DA) strategy. It consisted of 11 classic methods

(Flip, Scale, Rotation, Gaussian Blur, Gaussian Noise, Adjust Brightness, Adjust Contrast, Adjust Gamma, Cut Out, Sharpening), whereas some augmentations that nnU-Net uses were not applied (e.g., RemoveLabel, RenameTransform) [6]. This strategy was used for the IrMAs (class 1) as well as for the NV (class 3). For the NAs (class 2), on the other hand, we achieved better results with a more conservative augmentation strategy that included fewer methods (flip, scale, rotation).

Convolutions. Especially for the IrMAs, we encountered many false negatives, which we tried to tone down by increasing the number of convolutions per stage from two to one or two, respectively. This change did not directly lead to a significant improvement, but it was later used to create a combined prediction as described in the Sect. 4.3.

Model-Selection. The last changes in the nnU-Net pipeline were first, to not use the latest state of the model but rather the best according to the metrics calculated on the validation set and second, to build the default ensemble more selectively. For this ensemble, the framework used 5-fold cross-validation and built an ensemble from the corresponding five models [6]. Because of the already mentioned variance during training, we only included the best two to three models to avoid a lower overall Dice score based on the models with poor performance. This strategy consistently achieved higher DSC on the leaderboard than the standard configuration.

4.3 Postprocessing

As already mentioned in Sect. 4.2 we trained the class-specific models with different modifications, including the number of convolutions, data augmentation, model-selection as well as a dropout factor. This process resulted in three different approaches consisting of multiple runs, each with their respective predictions, which we used to generate our final submission. This was again done in a class-specific manner. For class 1, the problem was a large amount of undetected fine structures of the IrMAs, which resulted in a high False Negative Rate (FNR). Due to the modifications, each ensemble's prediction contained slight differences. To take advantage of this, the predictions for IrMAs were combined by using the union of all predictions. This resulted in a more stable prediction with reduced the variance caused during by the training. Since the NPAs and the NV were not suffering from the same FNR, we used a majority vote here. Even though the previous strategy works well for the IrMAs it decreased the DSC significantly for the other two classes. The reason for this was again the variance between training runs and the moderately high False Positive Rate (FPR) for NPAs and NV. Therefore the prediction did not benefit from the union over all prediction.

5   Related Tasks: Quality Assessment and DR Grading

Our team created an ensemble of different models to grade the quality of the available images and a similar ensemble for predicting the DR grading. The final submission contained the predictions of a ConvNext [7], EfficientNet V2 [11] and adapted Deep MIL [5] model for both tasks. For training, we split the provided training data further into our own training, validation, and test set. We tried ADAM and WADAM [8] without any learning rate scheduler as an optimizer, but our best models used ADAM only. As a loss function, we used Cross Entropy Loss and Masked BCE loss which was implemented with ideas from Unterthiner et al. [13]. To find the optimal hyperparameters for the individual models, we used a random search strategy.

Our approach for these tasks also included some modifications and changes compared to the standard implementations.

Data Augmentation. Our team used vertical flipping and moderate random crops to artificially increase the size of the provided data set before training our ConvNext and EfficientNet models.

Deep MIL. In contrast to the proposed feature extractor by Ilse et al. [5], we used our previously trained CNNs as feature extractors for the MIL approach. After finetuning the ConvNext and EfficientNet on the available training data, we used them to compute low-dimensional representations of the original image slices. To achieve that, we replaced the last classification layer of the CNNs with an identity function and saved the logits of the model in this stage. The size of the input slices was adapted to meet the model structure. For the final ensemble, only the MIL model with the ConvNext feature extractor was used.

Model Selection. Similar to task 1, we selected the hyperparameters with the highest performance (best kappa) on the validation set before retraining our models on the available training data.

6   Results

6.1   Task 1: Segmentation

Our team achieved third place in the segmentation task with a mean DSC of 0.5756. This result combined multiple methods containing a few training runs each. In the following table, the best results from each of the methods are shown as well as their combinations and the final submission (Table 2).

As already mentioned previously, we encountered the problem of inconsistency between training runs which is also reflected in the shown table as only the best results are presented. The average deviation for each of the classes was calculated to be 5%, 7%, and 6%, respectively. Unfortunately, we are not able to present scalable or promising postprocessing techniques for NPAs and NV. For IrMAs, however, the selected approach works well and achieves gradually better results for a larger amount of used models.

Table 2. Best results achieved with each model class and final submission. DA refers to the data augmentation that was previously proposed for each respective class.

Model	Class1	Class2	Class3
Baseline	0.4531	0.6463	0.5434
DA	0.4564	0.6592	**0.5917**
DA_Conv3/4	0.4559	0.6363	0.5688
Baseline + DA	0.4664	**0.6680**	0.5711
Baseline + DA + DA_Conv3/4	**0.4672**	0.6567	0.5754
Final Submit	0.4672	0.6680	0.5917

6.2   Task 2 and 3: Image Quality Assessment and Diabetic Retinopathy Grading

Our team placed 15th in the rankings for task 2, with our best submission achieving a quadratic weighted kappa of 0.7423. For task 3, we placed 12th with our best submission achieving a quadratic weighted kappa of 0.8240. These are the results of ensembles of different models for both tasks, as described previously. Unfortunately, our own hold-out test set was not suitable for getting a reliable estimate of our models on the public leaderboard. Therefore, we had to upload the predictions of several of our standalone models to get a better picture of our improvements. The final hyperparameters for our submitted models can be seen in Table 3 and Table 4 for tasks 2 and 3, respectively.

Table 3. Final hyperparameters for individual models of Task 2.

	Num. Epochs	Learning Rate	Weight Decay	Kappa
ConvNeXt_tiny	10	2e-5	0	0.63
EfficientNet V2	30	2e-4	0	0.65
Deep MIL	25	1e-3	0.7	0.69
Ensemble	-	-	-	**0.74**

Table 4. Final hyperparameters for individual models of Task 3.

	Num. Epochs	Learning Rate	Weight Decay	Kappa
ConvNeXt_tiny	30	6e-5	0.005	0.80
EfficientNet V2	10	9e-5	0	0.75
Deep MIL	10	6e-5	0	0.81
Ensemble	-	-	-	**0.82**

7 Conclusion

With the methods presented in Sect. 4, our team achieved third place on the leaderboard for the segmentation task. Despite the high variance during training, we achieved good results. We are especially confident in the postprocessing of the IrMAs. However, we would advise prioritizing the development of methods for stable and reproducible results, especially for the NPAs and the NV. This might lead to the conclusion that the stability can be traced back to the few samples with lower quality or might even hint at problems not yet known about this dataset.

Acknowledgements. The authors would like to thank the project AIRI FG 9-N (FWF-36284, FWF-36235) for helpful support. Further, the authors would like to thank ELLIS Unit Linz, the LIT AI Lab, and the Institute for Machine Learning for providing computing resources to participate in the challenge.

References

1. Dai, L., et al.: A deep learning system for detecting diabetic retinopathy across the disease spectrum. Nat. Commun. **12**(1), 1–11 (2021)
2. Esteva, A., et al.: Dermatologist-level classification of skin cancer with deep neural networks. Nature **542**(7639), 115–118 (2017)
3. ETDRS Research Group: Early treatment diabetic retinopathy study design and baseline patient characteristics: ETDRS report number 7. Ophthalmology **98**(5), 741–756 (1991)
4. Gulshan, V., et al.: Development and validation of a deep learning algorithm for detection of diabetic retinopathy in retinal fundus photographs. JAMA **316**(22), 2402–2410 (2016)
5. Ilse, M., Tomczak, J.M., Welling, M.: Attention-based deep multiple instance learning. In: 35th International Conference on Machine Learning (ICML), vol. 5, pp. 3376–3391 (2018)
6. Isensee, F., Jaeger, P.F., Kohl, S.A., Petersen, J., Maier-Hein, K.H.: nnU-Net: a self-configuring method for deep learning-based biomedical image segmentation. Nat. Methods **18**(2), 203–211 (2021)
7. Liu, Z., Mao, H., Wu, C.Y., Feichtenhofer, C., Darrell, T., Xie, S.: A convnet for the 2020s. In: Proceedings of the IEEE/CVF Conference on Computer Vision and Pattern Recognition, pp. 11976–11986 (2022)
8. Loshchilov, I., Hutter, F.: Decoupled weight decay regularization. arXiv preprint arXiv:1711.05101 (2017)
9. Radiuk, P.M.: Impact of training set batch size on the performance of convolutional neural networks for diverse datasets (2017)
10. Sheng, B., et al.: Diabetic retinopathy analysis challenge 2022 (2022). https://doi.org/10.5281/zenodo.6362349
11. Tan, M., Le, Q.: Efficientnetv2: smaller models and faster training. In: International Conference on Machine Learning, pp. 10096–10106. PMLR (2021)
12. Tian, M., Wolf, S., Munk, M.R., Schaal, K.B.: Evaluation of different swept' source optical coherence tomography angiography (SS-OCTA) slabs for the detection of features of diabetic retinopathy. Acta Ophthalmol. **98**(4), e416–e420 (2020)

13. Unterthiner, T., et al.: Deep learning for drug target prediction. In: Conference Neural Information Processing Systems Foundation (NeurIPS), Workshop on Representation and Learning Methods for Complex Outputs, vol. 2014 (2014)

14. Vislisel, J., Oetting, T.: Diabetic retinopathy: from one medical student to another. Eye Rounds (2010)

15. Zhang, Q., Rezaei, K.A., Saraf, S.S., Chu, Z., Wang, F., Wang, R.K.: Ultra-wide optical coherence tomography angiography in diabetic retinopathy. Quant. Imaging Med. Surg. 8(8), 743 (2018)

Automated Analysis of Diabetic Retinopathy Using Vessel Segmentation Maps as Inductive Bias

Linus Kreitner[1(✉)], Ivan Ezhov[1], Daniel Rueckert[1,2], Johannes C. Paetzold[2,3], and Martin J. Menten[1,2]

[1] Lab for AI in Medicine, Klinikum rechts der Isar, Technical University of Munich, Munich, Germany
linus.kreitner@tum.de
[2] BioMedIA, Department of Computing, Imperial College London, London, UK
[3] ITERM Institute Helmholtz Zentrum Muenchen, Neuherberg, Germany

Abstract. Recent studies suggest that early stages of diabetic retinopathy (DR) can be diagnosed by monitoring vascular changes in the deep vascular complex. In this work, we investigate a novel method for automated DR grading based on ultra-wide optical coherence tomography angiography (UW-OCTA) images. Our work combines OCTA scans with their vessel segmentations, which then serve as inputs to task specific networks for lesion segmentation, image quality assessment and DR grading. For this, we generate synthetic OCTA images to train a segmentation network that can be directly applied on real OCTA data. We test our approach on MICCAI 2022's DR analysis challenge (DRAC). In our experiments, the proposed method performs equally well as the baseline model.

Keywords: OCTA · Eye · Diabetic Retinopathy · MICCAI Challenges · Synthetic Data · Segmentation · Classification

1 Introduction

Optical coherence tomography angiography (OCTA) is a non-invasive *in-vivo* method to acquire high-resolution volumetric data from retinal microvasculature. These properties make OCTA efficient for identifying capillary abnormalities, which can often be observed in diabetic retinopathy. DR is a common medical condition that occurs as a result of untreated diabetes mellitus and causes 2.6% of all cases of blindness globally [1,2]. Roughly 11% of all diabetes patients develop vision-threatening complications such as neovascularization (NV), diabetic macular edema, or diabetic macular ischemia. Early stages of DR can be diagnosed, e.g., by non-perfusion areas (NAs) and intraretinal microvascular abnormalities (IRMAs).

J. C. Paetzold and M. J. Menten—Contributed equally as senior authors.

© The Author(s), under exclusive license to Springer Nature Switzerland AG 2023
B. Sheng and M. Aubreville (Eds.): MIDOG 2022/DRAC 2022, LNCS 13597, pp. 16–25, 2023.
https://doi.org/10.1007/978-3-031-33658-4_2

A recent survey by Sheng *et al.* calls OCT an "indispensable component of healthcare in ophthalmology", but notes that there are still many problems to be solved before artificial intelligence (AI) based systems can be reliably integrated in the clinical process [3]. The Diabetic Retinopathy Analysis Challenge (DRAC) is part of MICCAI 2022 in Singapore and aims at advancing machine learning research for DR analysis on OCTA data [4]. The organizers provide three labeled datasets with more than 1,000 ultra-wide swept-source OCTA (UW-OCTA) images. The samples are 1024×1024 pixel grayscale images and cover different field-of-views (FOVs). The challenge is divided into the tasks A) lesion segmentation, B) image quality assessment, and C) DR grading. Task A) is a pixelwise multi-label segmentation problem for identifying intraretinal microvascular abnormalities, non-perfusion areas, and neovascularization. The public training set contains 109 sample images with 86 cases of IRMA, 106 with NAs, and 35 with NVs. The training set of task B) contains 665 images, 77.9% of which are graded as excellent, 14.6% as good, and 7.5% as poor quality images. Finally, task C) contains 611 samples where 53.7% are classified as healthy, 34.9% as pre-proliferative DR, and the remaining 11.5% as proliferative DR (PDR). The DRAC event follows in the footsteps of similar challenges such as the Diabetic Retinopathy-Grading and Image Quality Estimation Challenge (DeepDRiD) for fundus images [5].

Diabetic retinopathy causes retinal vascular changes, and while being subtle at first, it eventually leads to dangerous vascular abnormalities and neovascularization. We hypothesize that by explicitly feeding the segmentation map of the vasculature to a deep neural network, we can provide a strong inductive bias for the network to predict the severity of DR. Recent studies suggest that early DR progression is mainly correlated with biomarkers found in the deep vascular complex (DVC), while the superficial vascular complex (SVC) is generally only affected in severe cases [1,6]. The authors hypothesize that the DVC might be more susceptible to ischemic damage because of its anatomical proximity to the outer plexiform layer, which has a high oxygen consumption. Therefore, it becomes clear that including small capillaries in the segmentation map is vital. However, to our knowledge, there is no publicly available dataset of OCTA images with corresponding vessel segmentation that includes small capillary vessels. Merely two public datasets, namely OCTA-500 and ROSE, have been released with segmentation labels for vessels and the foveal avascular zone (FAZ), both of which unfortunately only segment the largest vessels of the SVC [7,8]. To circumvent the problem of missing labeled data, we generate an artificial dataset ourselves. Figure 1 shows our proposed pipeline.

2   Methodology

In our work, we investigate whether an additional input channel to our network can boost the performance of a neural network. Our proposed additional input is a segmentation map of the vasculature, including small capillaries in the retina. To extract a faithful segmentation, we use synthetic OCTA images to train a segmentation network, which can then be used to generate the additional input channel on the fly.

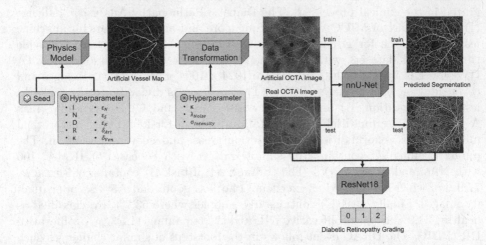

Fig. 1. The proposed framework for training a segmentation network using artificial training data. The pre-trained U-Net is then used to generate an additional segmentation layer channel for an arbitrary downstream task.

2.1   Synthetic Data Generation

Synthetic data generation to train segmentation networks is a vastly accepted technique to mitigate data sparsity through transfer learning in medical imaging [9–11]. Schneider *et al.* propose a configurable physics model to simulate the growth of vascular trees based on physical principles [12]. The blood vessel network is represented as a forest of rooted binary tree graphs that are iteratively updated based on the simulated oxygen concentration and the vascular endothelial growth factor (VEGF). Bifurcations and growth direction follow the physical laws of fluid dynamics to ensure realistic branching. From the resulting 3D graph network, it is then possible to retrieve 2D images and their ground truth segmentation maps by voxelizing the edges. This approach has proven efficient across multiple vessel segmentation tasks [13–16]. Menten *et al.* later adapted this approach to simulate the retinal vasculature [17]. They heuristically tune the simulator's hyperparameters to mimic the structure of the SVC and the DVC. Binary masks are used to decrease the VEGF within the FAZ, as well as the perimeter of the FOV. Using different seeds, it is possible to generate an unlimited amount of diverse vessel maps.

   To bridge the domain shift from synthetic to real OCTA images, Menten *et al.* employ a variety of data augmentations, such as eye motion artifacts, flow projection artifacts, and changes in brightness. The background signal caused by small capillary vessels is simulated by binomial noise and convoluted by a Gaussian filter since it is computationally intractable to simulate them. After training a U-Net on a synthetic dataset, the network can extract detailed segmentation maps on real data.

We adopt a modified algorithm version and tune it to realistically represent ultra-wide OCTA images. Instead of setting the roots of the trees on the image's border, we simulate the optical nerve, letting all vessel trees originate from there.

Instead of limiting bias fields to the outer rings of the FOV, we randomly apply them at every possible position as part of data augmentation. Furthermore, we employ random rotation, flipping, elastic transformation, scaling, and motion artifacts during the training of the segmentation network. We use a variation of the very successful U-Net architecture [18] and follow the guidelines of nnU-Net to configure our network optimally for the given task [19] (Fig. 2).

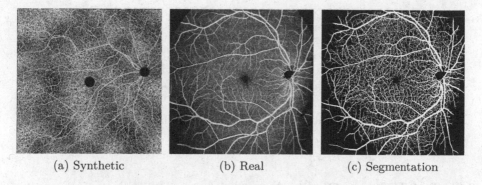

(a) Synthetic   (b) Real   (c) Segmentation

Fig. 2. Comparison of synthetic UW-OCTA images with real samples from the DRAC corpus [4]. The segmentation map was generated by a network trained only on synthetic data.

2.2   Task A: Lesion Segmentation

We define the segmentation task as a multi-label classification problem, where each pixel can be part of the classes IRMA, NA, NV, or neither. We choose the widely successful U-Net architecture, which is known for its reliable segmentation performance on medical image datasets [19,20]. For the optimal training parameters, we follow the guidelines of nnU-Net that define preprocessing steps, network architecture, loss function, and learning rate. Our final architecture can be seen in Fig. 3, where $C_{in} = 1$ for our baseline model, $C_{in} = 2$ for the network using the additional segmentation map input, and $C_{out} = 3$. Our loss function is the sum of the soft Dice loss and the channel-wise binary cross entropy loss.

Because of the small dataset and large class imbalances, we observe strong overfitting right from the start. We, therefore, apply extensive data augmentation, such as elastic deformation, contrast changes, Gaussian smoothing, flipping, and rotation. To further reduce overfitting, we initialize our proposed network with the weights from the pre-trained vessel segmentation network. We then train a baseline system as well as our proposed method for 300 epochs and select the checkpoint with the highest Dice similarity coefficient (DSC) on the validation set.

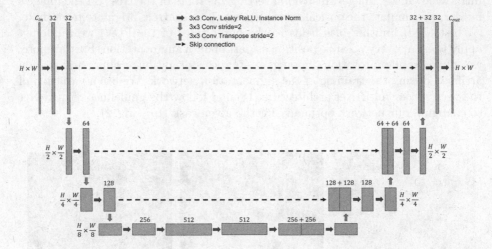

Fig. 3. Our U-Net architecture for Task A and for the vessel segmentation.

2.3 Task B: Image Quality Assessment

Task B is a multi-class classification problem, however, instead of using a classical cross-entropy loss, we redefine the objective as a regression problem using the Mean Squared Error (MSE) loss function. The rationale behind this is that the classes are not independent but rather share a cascading relationship with $C0 \prec C1 \prec C2$. Overfitting is a reoccurring pattern in all tasks. We find that larger models suffer more from overfitting than smaller models, leading us to test two widely known architectures, a ResNet18 and an EfficientNet_B0 [21,22]. Again, we compare the two baseline models with our proposed model, where the vessel segmentation map of an image is added as an additional input. We apply mild elastic deformation, flipping, 90° rotations, and contrast changes, as stronger augmentations may distort the characteristic appearance of the image artifacts. We train all models for 500 epochs and select the model with the best quadratic weighted kappa score.

2.4 Task C: Diabetic Retinopathy Grading

Our training method for DR grading is the same as for Task B. We increase the magnitude of the data augmentations and additionally add random rotations by $k \times 90° \pm 10°$ with $k \in \{0, 1, 2, 3\}$, Gaussian smoothing, and random erasing.

3   Results

During the challenge, we only tested the models performing best on a single validation set of each task. As the models all suffer from strong overfitting on the data split and biases by class imbalances, we decided to perform a stratified 5-fold cross-validation in the post challenge phase. We create an ensemble prediction on the test set using all trained models to obtain a more accurate estimation of the real performance. The official challenge test sets contain 65, 438, and 386 samples for tasks A, B, and C, respectively. Even though possible in the DRAC challenge, we specifically did not overfit on the test sets but instead only used the provided training set to select our models. The results for each task are listed in the following.

3.1   Results for Task A

To build the ensemble prediction for the segmentation task, we compute the pixel-wise and class-wise average score after applying a sigmoid activation. The output is then transformed into three distinct binary masks with a threshold $t = 0.5$. Despite our data augmentations, we observe overfitting. We could not find any significant difference between the baseline and the method using the additional vessel segmentation input. The results are shown in Table 1.

Table 1. Results for the lesion segmentation task. We computed the mean Dice Similarity Coefficient (mDSC) and the Intersection of Union (IoU) with their respective standard errors. In general, all methods perform similarly. The highest cross-validation scores are printed in bold.

Task A		(Cross-)Validation		Challenge Test Set	
		mDSC	IoU	mDSC	IoU
nnU-Net	Base+Seg	0.609	0.459	0.41	0.284
nnU-Net	Base	**0.558 ± 0.015**	**0.424 ± 0.015**	0.529 ± 0.011	0.385 ± 0.011
Ensemble	Base+Seg	0.556 ± 0.018	0.422 ± 0.018	0.518 ± 0.014	0.377 ± 0.013

3.2   Results for Task B

For the image classification task and the DR grading, the soft outputs from all trained models are averaged to form the ensemble prediction. First, comparing both architectures, we cannot find a significant performance gain from one over the other. During training, however, the EfficientNet was much more stable in its predictions, while the ResNet showed large jumps. Furthermore, the additional segmentation input did not seem to have meaningful benefits for training but instead caused the model to overfit faster. All results are listed in Table 2.

Table 2. Results for the image quality assessment task. We computed the Quadratic Weighted Kappa (QWK) score as well as the Area Under receiver operating characteristic Curve (AUC) with their respective standard errors. All methods perform similarly, with the EfficientNet being the most stable to train. The highest cross-validation scores are printed in bold.

Task B		(Cross-)Validation		Challenge Test Set	
		QWK	AUC	QWK	AUC
ResNet18	Base+Seg	0.917	0.931	0.677	0.802
ResNet18	Base	0.891 ± 0.011	$\mathbf{0.956 \pm 0.010}$	**0.726**	0.831
Ensemble	Base+Seg	0.891 ± 0.011	0.932 ± 0.009	0.699	**0.843**
EfficientNet_B0	Base	$\mathbf{0.901 \pm 0.010}$	0.930 ± 0.010	0.707	0.837
Ensemble	Base+Seg	0.893 ± 0.012	0.953 ± 0.010	0.662	0.818

3.3　Results for Task C

In the DR grading task, we employ the same approach as in Task B to compute the ensemble prediction. Again, we could not find a significant difference between the EfficientNet and the ResNet architecture, but the smoother training for the EfficientNet is also observed here. While the additional segmentation input did not yield any measurable benefit during cross-validation, the EfficientNet using the additional input outperformed the other models on the test set by a substantial margin. It is difficult to interpret, however, if this was caused by the additional information or due to random chance causing one version to perform better on the test set. Table 3 depicts all results in detail.

Table 3. Results for the diabetic retinopathy grading task. We computed the Quadratic Weighted Kappa score and the Area Under the receiver operating characteristic Curve with their respective standard errors. The EfficientNet seems to outperform the ResNet by a small margin but no substantial advantage for the additional segmentation input can be observed. The highest cross-validation scores are printed in bold.

Task C		(Cross-)Validation		Challenge Test Set	
		QWK	AUC	QWK	AUC
Resnet18	Base+Seg	0.838	0.894	0.277	0.666
Resnet18	Base	0.831 ± 0.014	0.911 ± 0.010	0.824	0.891
Ensemble	Base+Seg	0.846 ± 0.014	0.922 ± 0.010	0.805	0.891
EfficientNet_B0	Base	0.845 ± 0.010	$\mathbf{0.929 \pm 0.006}$	0.811	0.889
Ensemble	Base+Seg	$\mathbf{0.850 \pm 0.009}$	0.918 ± 0.004	**0.843**	**0.909**

4   Discussion

This study has investigated whether an additional channel containing the vessel segmentation map can boost the performance of a baseline model for the tasks of lesion segmentation, image quality assessment, and DR grading. This was motivated by the observation that early stages of DR manifest themselves in vascular changes of the DVC. We hypothesized that explicitly adding a vessel map as inductive bias would reduce overfitting through smaller segmentation and classification networks. While we were not able to win the DRAC challenge, we point out that this was not the primary goal of this study. Every team member was allowed to evaluate a model on the test set once per day, enabling teams with multiple members to test repeatedly and select their best model based on the test set. Additionally, the test set images were made public, making it possible to label the images manually. However, we deliberately abstained from extensive model tuning and overfitting on the test set.

Instead, we focused on testing the idea of integrating synthetic data into the prediction life cycle. We successfully extended Menten *et al.*'s work to generate OCTA images for ultra-wide scans. Using our training method, we can segment tiny capillaries whose diameter is similar to the physical resolution of the OCTA scanner. On this dataset, we could not find any significant improvement using the segmentation maps as input. However, we consider this finding relevant, as it indicates that the baseline models are strong enough to extract that information by themselves. It would be interesting to see whether this method could benefit from a larger image corpus, where networks are less prone to overfitting.

For future work, we aim to use the generated segmentation map to extract quantitative biomarkers. These in turn could be used to, e.g., train a random forest for DR grading. The prediction would hence become more explainable and allow doctors to retrace the network's reasoning.

5   Conclusion

This paper summarizes our contribution to the DR analysis challenge 2022. We present a novel method, where synthetic OCTA images are used to train a segmentation network, which then generates additional input channel for downstream tasks. Our method extracts state-of-the-art segmentation maps including tiny vessels from the deep vascular complex. However, we could not measure a substantial benefit of adding this auxiliary information to our deep learning pipeline in any of the three challenge tasks. Future work might build on our idea to use segmentation maps as complementary information and explore other applications where this is beneficial.

References

1. Sun, Z., Yang, D., Tang, Z., Ng, D.S., Cheung, C.Y.: Optical coherence tomography angiography in diabetic retinopathy: an updated review. Eye **35**(1), 149–161 (2021)

2. Flaxman, S.R., et al.: Global causes of blindness and distance vision impairment 1990–2020: a systematic review and meta-analysis. Lancet Global Health **5**(12), e1221–e1234 (2017)

3. Sheng, B., et al.: An overview of artificial intelligence in diabetic retinopathy and other ocular diseases. Front. Public Health **10**, 971943 (2022)

4. Sheng, B., et al.: Diabetic retinopathy analysis challenge (2022)

5. Liu, R., et al.: Deepdrid: diabetic retinopathy-grading and image quality estimation challenge. Patterns **3**(6), 100512 (2022)

6. Chua, J., et al.: Optical coherence tomography angiography in diabetes and diabetic retinopathy. J. Clin. Med. **9**(6), 1723 (2020)

7. Li, M., et al.: IPN-V2 and octa-500: methodology and dataset for retinal image segmentation (2020)

8. Ma, Y., et al.: Rose: a retinal oct-angiography vessel segmentation dataset and new model. IEEE Trans. Med. Imaging **40**(3), 928–939 (2021)

9. Raghu, M., Zhang, C., Kleinberg, J., Bengio, S.: Transfusion: understanding transfer learning for medical imaging. In: Advances in Neural Information Processing Systems, vol. 32 (2019)

10. Qasim, A.B., et al.: Red-GAN: attacking class imbalance via conditioned generation. Yet another medical imaging perspective. In: Medical Imaging with Deep Learning, pp. 655–668. PMLR (2020)

11. Horvath, I., et al.: Metgan: generative tumour inpainting and modality synthesis in light sheet microscopy. In: Proceedings of the IEEE/CVF Winter Conference on Applications of Computer Vision, pp. 227–237 (2022)

12. Schneider, M., Reichold, J., Weber, B., Székely, G., Hirsch, S.: Tissue metabolism driven arterial tree generation. Med. Image Anal. **16**(7), 1397–1414 (2012)

13. Gerl, S., et al.: A distance-based loss for smooth and continuous skin layer segmentation in optoacoustic images. In: Martel, A.L., et al. (eds.) MICCAI 2020. LNCS, vol. 12266, pp. 309–319. Springer, Cham (2020). https://doi.org/10.1007/978-3-030-59725-2_30

14. Shit, S., et al.: clDice-a novel topology-preserving loss function for tubular structure segmentation. In: Proceedings of the IEEE/CVF Conference on Computer Vision and Pattern Recognition, pp. 16560–16569 (2021)

15. Todorov, M.I., et al.: Machine learning analysis of whole mouse brain vasculature. Nat. Methods **17**(4), 442–449 (2020)

16. Paetzold, J.C., et al.: Whole brain vessel graphs: a dataset and benchmark for graph learning and neuroscience. In: Thirty-Fifth Conference on Neural Information Processing Systems Datasets and Benchmarks Track (Round 2) (2021)

17. Menten, M.J., Paetzold, J.C., Dima, A., Menze, B.H., Knier, B., Rueckert, D.: Physiology-based simulation of the retinal vasculature enables annotation-free segmentation of oct angiographs. In: Wang, L., Dou, Q., Fletcher, T., Speidel, S., Li, S. (eds.) MICCAI 2022. LNCS, vol. 13438, pp. 330–340. Springer, Cham (2022). https://doi.org/10.1007/978-3-031-16452-1_32

18. Ronneberger, O., Fischer, P., Brox, T.: U-Net: convolutional networks for biomedical image segmentation. In: Navab, N., Hornegger, J., Wells, W.M., Frangi, A.F. (eds.) MICCAI 2015. LNCS, vol. 9351, pp. 234–241. Springer, Cham (2015). https://doi.org/10.1007/978-3-319-24574-4_28

19. Isensee, F., Jaeger, P.F., Kohl, S.A.A., Petersen, J., Maier-Hein, K.H.: nnU-Net: a self-configuring method for deep learning-based biomedical image segmentation. Nat. Methods **18**(2), 203–211 (2021)

20. Çiçek, Ö., Abdulkadir, A., Lienkamp, S.S., Brox, T., Ronneberger, O.: 3D U-Net: learning dense volumetric segmentation from sparse annotation. In: Ourselin, S., Joskowicz, L., Sabuncu, M.R., Unal, G., Wells, W. (eds.) MICCAI 2016. LNCS, vol. 9901, pp. 424–432. Springer, Cham (2016). https://doi.org/10.1007/978-3-319-46723-8_49
21. He, K., Zhang, X., Ren, S., Sun, J.: Deep residual learning for image recognition. In: 2016 IEEE Conference on Computer Vision and Pattern Recognition (CVPR), pp. 770–778 (2016)
22. Tan, M., Le, Q.: Efficientnet: rethinking model scaling for convolutional neural networks. In: International Conference on Machine Learning (2019)

Bag of Tricks for Diabetic Retinopathy Grading of Ultra-Wide Optical Coherence Tomography Angiography Images

Renyu Li[1($\boxtimes$)], Yunchao Gu[1], Xinliang Wang[1], and Sixu Lu[2]

[1] Beihang University, Beijing, China
LPudding@buaa.edu.cn
[2] Beijing Normal University, Beijing, China

Abstract. The performance of disease classification can be improved through improvements in the training process, such as changes in data augmentation, optimization methods, and deep learning model architectures. In the Diabetic Retinopathy Analysis Challenge, we employ a series of techniques to enhance the performance of the diabetic retinopathy grading. In this paper, we examine a collection of these improvements and empirically evaluate their impact on the final model accuracy through experiments. Experiments show that these improvements can significantly improve the performance of the model. For this task, we use a single SeResNext to improve the validation score from 0.8322 to 0.8721.

Keywords: DRAC · Diabetic retinopathy grading · Medical image analysis · Deep learning

1 Introduction

Diabetic retinopathy (DR) is one of the leading causes of blindness, affecting approximately 78% of people with a history of diabetes of 15 years or more [9]. DR often causes gradual changes in the structure of the vasculature. DR is diagnosed by the presence of retinopathy such as microaneurysms, intraretinal microvascular abnormalities, non-perfused areas, and neovascularization. The detection of these lesions is crucial for the diagnosis and treatment of DR.

With the development of deep learning technology, a series of automatic DR diagnosis methods based on deep learning have been proposed, which can help doctors to specify tailor-made treatments for patients [3,8,12–14,16]. Currently, there have been some works using fundus images for DR diagnosis. [14] used the attention map to highlight suspicious areas while completing DR grading and lesion localization. [16] proposed a collaborative learning approach for semi-supervised lesion segmentation and disease grading. [13] formulated lesion identification as a weakly supervised lesion localization problem via a transformer decoder that jointly performs DR grading and lesion detection.

© The Author(s), under exclusive license to Springer Nature Switzerland AG 2023
B. Sheng and M. Aubreville (Eds.): MIDOG 2022/DRAC 2022, LNCS 13597, pp. 26–30, 2023.
https://doi.org/10.1007/978-3-031-33658-4_3

However, these traditional works use fundus photography or fundus fluorescein angiography(FFA) for diagnosis. Fundus photography is difficult to detect early or small neovascular lesions. FA is an invasive fundus imaging method and cannot be used in patients with allergies, pregnancy, or poor liver and kidney function. Ultra-wide optical coherence tomography angiography(UW-OCTA) imaging can non-invasively detect the changes of DR neovascularization and is an important imaging modality. However, there are currently no works capable of automatic DR analysis using UW-OCTA. In the Diabetic Retinopathy Analysis Challenge, the use of UW-OCTA for DR grading is an important challenge for automatic DR analysis. In this paper, we introduce a set of tricks we use in the competition and demonstrate the effectiveness of our method through extensive experiments.

2   Bag of Tricks

2.1   Data Augmentation

Image augmentation is a method commonly used in computer vision tasks to improve the quality of trained models. It improves the generalization of the model by creating new training samples from existing data. We use the Albumentations [2] which is a commonly used Python library for data augmentation in this challenge. The data augmentation methods we use include Resize, HorizontalFlip, RandomBrightnessContrast, ShiftScaleRotate, CLAHE, and Cutout. In particular, Cutout [4] removes some regions in the image, forcing the model to focus on other regions to complete the classification task, which can greatly reduce the possibility of overfitting. CLAHE [11] transforms the histogram distribution of the image into an approximately uniform distribution, facilitating training by enhancing the contrast of the image.

. In addition, we use the mixup [15] technique during training. Mixup expands the training dataset by mixing images of different categories, enhancing the generalization of model training. We use the test-time augmentation method during testing, returning the set of these predictions by data augmentation methods such as flipping, and rotating, to achieve better predictions.

2.2   Optimizer

Optimization techniques play a crucial role in the training of deep neural networks, which can help models converge to better positions faster and more stably. The common optimizers include SGD [1], Adam [7] and so on. In this competition, we use the Adam optimizer for training. Adam optimizer absorbs the advantages of Adagrad [5] and momentum gradient descent algorithms [10], which can not only adapt to sparse gradients but also alleviate the problem of gradient oscillation.

Table 1. Quantitative evaluation results. DA, mixup, smoothing, CALR, and tta respectively denote the application of data augmentation, mixup technique, label smoothing, CosineAnnealingLR strategy, and test-time augmentation. CV represents the result of cross-validation, and LB represents the result on the leader board.

DA	mixup	smoothing	CALR	tta	CV	LB
					0.8177	0.8322
✓					0.8188	0.8510
✓	✓				0.8167	0.8610
✓	✓	✓			0.8202	0.8665
✓	✓	✓	✓		0.8199	0.8690
✓	✓	✓	✓	✓	**0.8206**	**0.8721**

2.3 Learning Rate Scheduler

The learning rate is the most important parameter in training a neural network. A too-high learning rate will speed up the learning in the early stage of algorithm optimization, making it easier for the model to approach the local or global optimal solution. However, there will be large fluctuations in the later stage, and even the value of the loss function is hovering around the minimum value, and it is difficult to reach the optimal value. We use the strategy of CosineAnnealingLR to adjust the learning rate during training. Using the cosine function as the period, it resets the learning rate to the maximum value for each period, preventing the training process from converging to a local minimum.

2.4 Loss Function

The Cross Entropy loss is the most commonly used classification loss function. In general, labels are usually encoded as one-hot codes, and this type of hard label is sensitive to noisy labels and easily leads to overfitting. Soft labels can bring stronger generalization capabilities to the model and are relatively robust to hierarchical data and noisy labels. Therefore, we use label smoothing to assign a small part of the probability to other classes, so that the labels are not so absolute, which can effectively prevent overfitting.

3 Experiments

3.1 Dataset and Evaluation Metrics

Our dataset contains 997 images, of which 611 images are used for training and 386 images are used for testing. This dataset is provided by the Diabetic Retinopathy Grading Challenge, which contains 3 grades. Quadratic weighted kappa is used to evaluate the performance of grading.

3.2 Implementation Details

We conduct experiments using the techniques described above as well as a commonly used CNN model SeResNext [6]. For convenience, our models are all performed with an image size of 512*512. Since the dataset is small, we use k-fold cross-validation for training set partitioning, where k is 4. With a learning rate of 5e-4 and a batch size of 32, all our models are trained for 25 epochs using the Adam optimizer and CosineAnnealingLR.

3.3 Results

Table 1 shows the quantitative evaluation results of our experiments. It can be found that data augmentation and mixup greatly improve the performance of the model by generating new samples from existing data. In addition, label smoothing and CosineAnnealingLR strategy improve the model performance to some extent by smoothing the training. Finally, test-time augmentation improves the robustness of the model through test data augmentation, which further improves the performance of the model.

4 Conclusions

In this paper, we demonstrate the techniques we used in the Diabetic Retinopathy Analysis Challenge. These improvements greatly improve the performance of the model, which can help doctors make diagnoses. In the future, we will use additional UW-OCTA datasets for pre-training to further improve the performance of the model.

References

1. Bottou, L.: Large-scale machine learning with stochastic gradient descent. In: Lechevallier, Y., Saporta, G. (eds.) COMPSTAT 2010, pp. 177–186. Springer, Heidelberg (2010). https://doi.org/10.1007/978-3-7908-2604-3_16
2. Buslaev, A., Iglovikov, V.I., Khvedchenya, E., Parinov, A., Druzhinin, M., Kalinin, A.A.: Albumentations: fast and flexible image augmentations. Information 11(2), 125 (2020)
3. Dai, L., et al.: A deep learning system for detecting diabetic retinopathy across the disease spectrum. Nat. Commun. 12(1), 1–11 (2021)
4. DeVries, T., Taylor, G.W.: Improved regularization of convolutional neural networks with cutout. arXiv preprint arXiv:1708.04552 (2017)
5. Duchi, J., Hazan, E., Singer, Y.: Adaptive subgradient methods for online learning and stochastic optimization. J. Mach. Learn. Res. 12(7), 2121–2159 (2011)
6. Hu, J., Shen, L., Sun, G.: Squeeze-and-excitation networks. In: Proceedings of the IEEE Conference on Computer Vision and Pattern Recognition, pp. 7132–7141 (2018)
7. Kingma, D.P., Ba, J.: Adam: a method for stochastic optimization. arXiv preprint arXiv:1412.6980 (2014)

8. Liu, R., et al.: Deepdrid: diabetic retinopathy-grading and image quality estimation challenge. Patterns **3**, 100512 (2022)
9. Na, K.I., Lee, W.J., Kim, Y.K., Jin, W.J., Park, K.H.: Evaluation of optic nerve head and peripapillary choroidal vasculature using swept-source optical coherence tomography angiography. J. Glaucoma **26**(7), 665 (2017)
10. Nesterov, Y.: A method for unconstrained convex minimization problem with the rate of convergence (1983)
11. Reza, A.M.: Realization of the contrast limited adaptive histogram equalization (CLAHE) for real-time image enhancement. J. VLSI Signal Process. Syst. Signal Image Video Technol. **38**(1), 35–44 (2004)
12. Sheng, B., et al.: An overview of artificial intelligence in diabetic retinopathy and other ocular diseases. Front. Public Health **10** (2022)
13. Sun, R., Li, Y., Zhang, T., Mao, Z., Wu, F., Zhang, Y.: Lesion-aware transformers for diabetic retinopathy grading. In: Proceedings of the IEEE/CVF Conference on Computer Vision and Pattern Recognition, pp. 10938–10947 (2021)
14. Wang, Z., Yin, Y., Shi, J., Fang, W., Li, H., Wang, X.: Zoom-in-net: deep mining lesions for diabetic retinopathy detection. In: Descoteaux, M., Maier-Hein, L., Franz, A., Jannin, P., Collins, D.L., Duchesne, S. (eds.) MICCAI 2017. LNCS, vol. 10435, pp. 267–275. Springer, Cham (2017). https://doi.org/10.1007/978-3-319-66179-7_31
15. Zhang, H., Cisse, M., Dauphin, Y.N., Lopez-Paz, D.: mixup: beyond empirical risk minimization. arXiv preprint arXiv:1710.09412 (2017)
16. Zhou, Y., He, X., Huang, L., Liu, L., Zhu, F., Cui, S., Shao, L.: Collaborative learning of semi-supervised segmentation and classification for medical images. In: Proceedings of the IEEE/CVF Conference on Computer Vision and Pattern Recognition, pp. 2079–2088 (2019)

Deep Convolutional Neural Network
for Image Quality Assessment
and Diabetic Retinopathy Grading

Zhenyu Chen(ID) and Liqin Huang(✉)(ID)

College of Physics and Information Engineering, Fuzhou University, Fuzhou, China
hlq@fzu.edu.cn

Abstract. Quality assessment of ultra-wide optical coherence tomography angiography (UW-OCTA) images followed by lesion segmentation and proliferatived diabetic retinopathy (PDR) detection is of great significance for the diagnosis of diabetic retinopathy. However, due to the complexity of UW-OCTA images, it is challenging to achieve automatic image quality assessment and PDR detection in a limited dataset. This work presented a fully automated convolutional neural network-based method for image quality assessment and retinopathy grading. In the first stage, the dataset was augmented to eliminate the category imbalance problem. In the second stage, the EfficientNet-B2 network, pre-trained on ImageNet, was used for quality assessment and lesion grading of UW-OCTA images. We evaluated our method on the DRAC2022 dataset. A quadratic weighted kappa score of 0.7704 was obtained on the task 2 image quality assessment test set and 0.8029 on the task 3 retinopathy grading test set.

Keywords: Image quality assessment · Diabetic retinopathy grading · Convolution neural network

1 Introduction

Diabetic retinopathy (DR) is an eye disease caused by high blood glucose levels that can cause irreversible visual impairment and blindness [14]. Accurate grading of diabetic retinopathy is a time-consuming task for ophthalmologists and a major challenge for beginners in ophthalmology. Ultra-wide optical coherence tomography angiography (UW-OCTA) can detect changes in DR neovascularization noninvasively and is an important imaging modality to help ophthalmologists diagnose PDR. However, there are no works that enable automated DR analysis using UW-OCTA. Using computer-aided methods to assess the image quality of UW-OCTA and then achieve DR grading is of great significance for the diagnosis of diabetic retinopathy.

In the traditional retinal image analysis methods, most of the implementation is to extract the image brightness, color, shape, and other features, and

B. Sheng and M. Aubreville (Eds.): MIDOG 2022/DRAC 2022, LNCS 13597, pp. 31–37, 2023.
https://doi.org/10.1007/978-3-031-33658-4_4

then use the classification algorithm to analyze. Nayak et al. [10] used machine learning for feature extraction to classify only proliferative diabetic retinopathy and non-proliferative diabetic retinopathy, Mookiah et al. [9] used AdaBoost method to extract features for DR classification. Shahin et al. [11] used morphological processing methods to extract pathological features such as vascular area and exudation area, and obtained 0.88 sensitivity and 1.0 specificity in classifying diabetic retinopathy. Casanova et al. [2] used the random forest algorithm for DR classification, and assessed DR risk according to graded fundus photos and system data, with an accuracy rate of 0.9. However, these methods rely on manual annotation by clinical ophthalmologists and require significant effort to manually extract features using image processing methods, thus introducing additional complexity and instability to the classification task.

In recent years, deep learning has achieved remarkable results on DR image classification tasks [8, 12]. Dai L et al. [3] developed a deep learning system for the detection of diabetic retinopathy. Gulshan et al. [5] used the Inception V3 deep model to detect diabetic retinopathy using data sets of more than 128000 fundus images, and achieved very high AUC values in two different test sets. Shaohua Wana et al. [15] used AlexNet, VggNet, GoogleNet, and ResNet152 as four network models in DR image classification, the best classification accuracy was the VggNet-s model with 0.9568 accuracy. Rishab et al. [4] proposed a method that combines deep CNN with traditional machine learning algorithms, and used residual network and decision tree classifier to classify DR. The model was verified on the MESSIDOR-2 dataset. Mammoth team [16] used the DenseNet121 network for feature extraction and machine learning boosted tree algorithm for prediction, which was tested on the dataset and verified that the model has better performance. In 2019 Philippe et al. [1] used GAN to artificially synthesize retinal images to expand the dataset and improve the ability of training DR classification models. At present, deep learning was mainly applied to DR analysis of OCTA images, and the method of automatic DR analysis using UW-OCTA is still to be developed.

In this work, we developed a fully automated method for image quality assessment and retinopathy grading based on convolutional neural networks. The main contributions of our method are summarized as follows.

1. We used the EfficientNet-B2 network pre-trained on ImageNet, trained and evaluated on the DRAC2022 dataset.
2. We adopted a data augmentation strategy to cope with the imbalance in the number of categories in the challenge dataset.
3. We used five-fold cross-validation to evaluate the model performance to adjust the hyperparameters of the neural network.

2   Methods

Figure 1 shows the method used for UW-OCTA image quality assessment and retinopathy grading. In the first stage, the original image was augmented. In the second stage, the expanded images were uniformly adjusted to 384 × 384 and input to the network for image quality assessment and lesion grading.

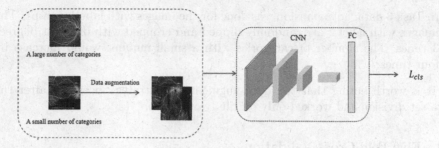

Fig. 1. Image quality assessment and retinopathy grading framework.

2.1 Backbone

We adopted three classical convolutional neural networks ResNet50 [6], Dense-Net121 [7], and EfficientNet-B2 [13] as candidate backbone networks. ResNet solved the problem of gradient disappearance and explosion in deep networks by introducing residual structure into the network. DenseNet through intensive connection relieved the problem depth network disappeared, strengthened the characteristics of transmission, and largely reduced the number of parameters. EfficientNet used composite coefficients to uniformly scale all dimensions of the model, which greatly improved the accuracy and efficiency of the model. In order to speed up the model training and avoid gradient disappearance or gradient explosion, the candidate backbone networks all adopted the pre-trained weight parameters of ImageNet.

The loss function used is CrossEntropyLoss:

$$CrossEntropyLoss(g, p) = -\sum_{i=1}^{C} g_i \log(p_i) \tag{1}$$

where C denotes the number of classes, g_i is label, p_i is prediction.

2.2 Data Augmentation

Before training, we performed category evaluation on the dataset of DRAC2022 task 2 and task 3. We found that the dataset has a serious category imbalance problem:

1) In task 2 data, images with label 2 account for 78% of all images.
2) In task 3 data, images with label 2 account for 8% of all images.

To solve the category imbalance problem of the data, we used the following augmentation strategy:

1) In task 2 data, no augmentation was done for the images with label 2. The remaining labeled images were randomly flipped and cropped with different degrees 3 times. The number of categories with a small number would increase by four times.

2) In Task 3 data, no expansion was done for the images with labels 0 and 1. The images with label 2 are randomly flipped and cropped with different degrees 3 times. The number of categories with a small number would increase by four times.

It is worth noting that the data augmentation strategy occurred after the data set division and worked only on the training set.

2.3 Five Fold Cross Validation

As shown in Fig. 2, we used five-fold cross-validation to adjust the hyperparameters of our model in the following steps.

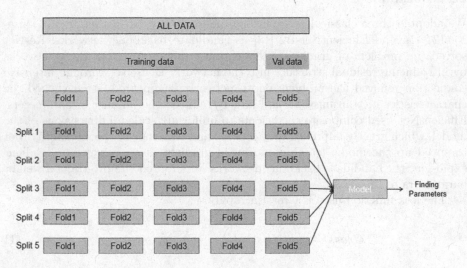

Fig. 2. Five fold cross validation.

1. The entire dataset was divided into the Training set and Validation set according to the same proportion of categories: specifically, we divided all the datasets into five parts and selected four parts each time as the Training set and performed data augmentation, and one part as the Validation set.
2. Adjust the parameters for different models: select a candidate model, cross-validate it with different hyperparameters, and got the evaluation result; selected the hyperparameter with the best evaluation result as the optimal hyperparameter; train the model again with the Training set for the optimal hyperparameter.
3. Evaluate the generalization ability of different models using the official test data provided.

3   Experiment

3.1   Dataset

DRAC2022 provided a standardized UW-OCTA data set for image quality assessment and DR grading.

1) Image quality evaluation data set: The training set consists of 665 images and corresponding labels in CSV files. The dataset contains three different image quality levels: poor quality level (0), good quality level (1), and excellent quality level (2).
2) Diabetic retinopathy grading data set: The training set consists of 611 images and corresponding labels in CSV files. This dataset contains three different grades of diabetic retinopathy: Normal (0), NPDR (1), and PDR (2).

NPDR: non-proliferatived diabetic retinopathy; PDR: proliferatived diabetic retinopathy.

3.2   Implementations

When training the model, we resized each image to 384×384 to reduce GPU memory consumption and training time. We choose PyTorch to implement our model and use an NVIDIA GeForce RTX 3080 GPU for training. The input size of the network is 384×384 and the batch size is 8; in our model, we set the epoch to 40, the initial learning rate to 1×10^{-2}., and adopted the Adam optimizer to minimize Loss to update the network parameters.

3.3   Results

We used quadratic weighted kappa as an evaluation metric to assess the performance of our framework. We evaluated the performance of several networks, including ResNet50, DenseNet121, and EfficientNet-B2. In addition, to verify the effectiveness of the data augmentation strategy, we also performed ablation experiments on candidate networks.

Table 1 shows the results of image quality evaluation, among the three candidate networks, EfficienNet-B2 achieved better performance compared to ResNet50, DenseNet121. It can also be seen that our proposed data augmentation strategy achieved better results in the category imbalanced dataset. With the same backbone network, quadratic weighted kappa achieves at least 0.02 improvement. Especially when the performance of the network itself is poor, the improvement can be up to 0.05.

Table 2 shows the results for diabetic retinopathy grading, and among the three candidate networks, EfficientNet-B2 achieved better performance compared to the other series of networks. Our proposed data augmentation strategy achieved better results in the category imbalanced dataset, with good improvement in quadratic weighted kappa with the same backbone network. On ResNet50, the improvement is 0.025, on DenseNet121, 0.044, and on EfficientNet-B2, 0.028.

Table 1. Quadratic Weight Kappa for image quality assessment.

Method	Data Augmentation	Quadratic Weight Kappa
ResNet50	×	0.6362
ResNet50	✓	0.6912
DenseNet121	×	0.7025
DenseNet121	✓	0.7353
EfficientNet-B2	×	0.7457
EfficientNet-B2	✓	**0.7704**

Table 2. Quadratic Weight Kappa for diabetic retinopathy grading.

Method	Data Augmentation	Quadratic Weight Kappa
ResNet50	×	0.6815
ResNet50	✓	0.7173
DenseNet121	×	0.7247
DenseNet121	✓	0.7688
EfficientNet-B2	×	0.7703
EfficientNet-B2	✓	**0.8029**

4   Conclusion

In this work, we developed a fully automated image quality assessment and lesion grading method based on a pre-trained EfficientNet-B2 network. We employed a data augmentation strategy to address the class imbalance of the training set. We used a five-fold cross-validation to evaluate the model performance as a way to adjust the hyperparameters of the neural network. The method was evaluated on the DRAC2022 dataset. The experimental results show that the method achieves good results in image quality assessment and lesion grading tasks.

References

1. Burlina, P.M., Joshi, N., Pacheco, K.D., Liu, T.A., Bressler, N.M.: Assessment of deep generative models for high-resolution synthetic retinal image generation of age-related macular degeneration. JAMA Ophthalmol. **137**(3), 258–264 (2019)
2. Casanova, R., Saldana, S., Chew, E.Y., Danis, R.P., Greven, C.M., Ambrosius, W.T.: Application of random forests methods to diabetic retinopathy classification analyses. PLoS ONE **9**(6), e98587 (2014)
3. Dai, L., et al.: A deep learning system for detecting diabetic retinopathy across the disease spectrum. Nat. Commun. **12**(1), 1–11 (2021)
4. Gargeya, R., Leng, T.: Automated identification of diabetic retinopathy using deep learning. Ophthalmology **124**(7), 962–969 (2017)

5. Gulshan, V., et al.: Development and validation of a deep learning algorithm for detection of diabetic retinopathy in retinal fundus photographs. JAMA **316**(22), 2402–2410 (2016)
6. He, K., Zhang, X., Ren, S., Sun, J.: Deep residual learning for image recognition. In: Proceedings of the IEEE Conference on Computer Vision and Pattern Recognition, pp. 770–778 (2016)
7. Huang, G., Liu, Z., Van Der Maaten, L., Weinberger, K.Q.: Densely connected convolutional networks. In: Proceedings of the IEEE Conference on Computer Vision and Pattern Recognition, pp. 4700–4708 (2017)
8. Liu, R., et al.: Deepdrid: diabetic retinopathy-grading and image quality estimation challenge. Patterns 100512 (2022)
9. Mookiah, M.R.K., Chua, C.K., Min, L.C., Ng, E., Laude, A.: Computer aided diagnosis of diabetic retinopathy using multi-resolution analysis and feature ranking frame work. J. Med. Imaging Health Inform. **3**(4), 598–606 (2013)
10. Nayak, J., Bhat, P.S., Acharya, U., Lim, C.M., Kagathi, M., et al.: Automated identification of diabetic retinopathy stages using digital fundus images. J. Med. Syst. **32**(2), 107–115 (2008)
11. Shahin, E.M., Taha, T.E., Al-Nuaimy, W., El Rabaie, S., Zahran, O.F., Abd El-Samie, F.E.: Automated detection of diabetic retinopathy in blurred digital fundus images. In: 2012 8th International Computer Engineering Conference (ICENCO), pp. 20–25. IEEE (2012)
12. Sheng, B., et al.: An overview of artificial intelligence in diabetic retinopathy and other ocular diseases. Front. Public Health **10** (2022)
13. Tan, M., Le, Q.: Efficientnet: rethinking model scaling for convolutional neural networks. In: International Conference on Machine Learning, pp. 6105–6114. PMLR (2019)
14. Tian, M., Wolf, S., Munk, M.R., Schaal, K.B.: Evaluation of different swept'source optical coherence tomography angiography (SS-OCTA) slabs for the detection of features of diabetic retinopathy. Acta Ophthalmol. **98**(4), e416–e420 (2020)
15. Wan, S., Liang, Y., Zhang, Y.: Deep convolutional neural networks for diabetic retinopathy detection by image classification. Comput. Electr. Eng. **72**, 274–282 (2018)
16. Wang, Yu., Wang, G.A., Fan, W., Li, J.: A deep learning based pipeline for image grading of diabetic retinopathy. In: Chen, H., Fang, Q., Zeng, D., Wu, J. (eds.) ICSH 2018. LNCS, vol. 10983, pp. 240–248. Springer, Cham (2018). https://doi.org/10.1007/978-3-030-03649-2_24

Diabetic Retinal Overlap Lesion Segmentation Network

Zhiqiang Gao and Jinquan Guo[✉]

School of Mechanical Engineering and Automation, Fuzhou University, Fuzhou, China
maxguo@163.com

Abstract. Diabetic retinopathy(DR) is the major cause of blindness, and the pathogenesis is unknown. Ultra-wide optical coherence tomography angiography imaging (UW-OCTA) can help ophthalmologists to diagnose DR. Automatic and accurate segmentation of lesions is essential for the diagnosis of DR, yet accurate identification and segmentation of lesions from UW-OCTA images remains a challenge. We proposed a modified nnUNet named nnUNet-CBAM. Three networks were trained to segment each lesion separately. Our method was evaluated in DRAC2022 diabetic retinopathy analysis challenge, where segmentation results were tested on 65 UW-OCTA images. These images are standardized UW-OCTA. Our method achieved a mean dice similarity coefficient (mDSC) of 0.4963 and a mean intersection over union (mIOU) of 0.3693.

Keywords: Diabetic Retinopathy · Deep Learning · Medical Segmentation

1 Introduction

Diabetic retinopathy (DR) is a common eye disease in diabetic patients and is a significant cause of blindness in the population, with approximately 78% of the population have diabetes and have a medical history of the disease for more than 15 years [15]. The exact mechanism of diabetic retinopathy is unknown, but it is generally believed to be related to retinal microvascular damage. When human blood sugar is too high, it will cause the thickness of microvascular basement membrane, which will lead to a decrease in the caliber of the vessels, the roughness of the inner wall, and a decrease in elasticity and contractility; then, the microvasculature distributed in the retina will be very vulnerable, the phenomenon may lead to rupture and hemorrhage the microvasculature because of its fragility [1]. DR often causes gradual changes in the structure of the vascular system and results in abnormalities. DR is diagnosed by visual inspection of images to check for the presence of retinal lesions, such as microaneurysms (MAs), intraretinal microvascular abnormalities (IRMAs), nonperfused areas (NPAs), and neovascularization(NV). The detection of these lesions is critical to the diagnosis of DR.

B. Sheng and M. Aubreville (Eds.): MIDOG 2022/DRAC 2022, LNCS 13597, pp. 38–45, 2023.
https://doi.org/10.1007/978-3-031-33658-4_5

In recent years, optical coherence tomography angiography(OCTA) has been extensively used in the study of ophthalmic diseases, it has significant advantages in the sensitivity and accuracy of detecting related diseases, and has a very promising application. Swept-source (SS)-OCTA allows individual assessment of the choroidal vascular system. Several works have been performed to grade the qualitative features of diabetic retinopathy [7, 12, 14]. Some works already use Ultra-wide optical coherence tomography angiography(UW-OCTA) on DR analysis [10, 16]. However, there are currently no works that can use UW-OCTA for automatic DR analysis, and the automatic segmentation of fundus lesions from UW-OCTA still has the following problems. Firstly, as shown in Fig. 1(a), there remain imaging defects in the boundaries of some images. Secondly, there exist motion artifacts in UW-OCTA imaging, which are manifested as horizontal or vertical dark or bright streaks, image misalignment, stretching, or distortion as reflected in Fig. 1(b).

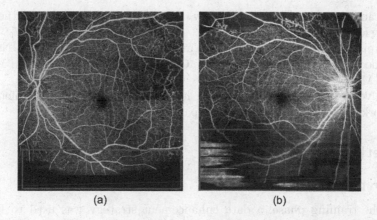

(a)   (b)

Fig. 1. (a) Boundary defects (b) Motion artifacts.

The basic process of traditional pattern recognition methods is to extract the features of the microvascular tumor region firstly, and then classify the features using Random Forest (RF) or Support Vector Machines (SVM). A statistical method based on hybrid model clustering and logistic regression was proposed in the literature [11], the method uses a hybrid model to fit the grayscale distribution of retinal images, divides the pixels in the image into the foreground (microangioma) and background (optic disc, blood vessels, etc.), then classifies the candidate targets by using the foreground as a candidate region and extracting features from it, a logical regression classifier was used to classify the candidate targets finally.

In the past few years, there has been an increasing interest in automatic DR recognition using deep learning techniques based on convolutional neural networks (CNN) [13]. Comparing the recognition performance of CNN networks with human experts, the conclusion is that deep neural networks can compete with human experts in DR recognition performance [2–4]. Kou et al [6] proposed Deep Recurrent UNet (DRUNet), which combined the deep residual model and recursive convolution into the UNet structure, and its area under the curve(AUC) values reaches 0.994 and 0.987 on E-Ophtha and IDRiD datasets. Compared to the original UNet network with 0.838 and 0.761 on the above two datasets, the DRUNet has a considerable improvement in segmentation results. But the following problems still exist for lesion segmentation:

1. The number of cases with different lesions is not similar or equal.
2. The segmentation result of overlapping lesions is not ideal.
3. The class imbalance between foreground and background is more serious.

To address the above problem, we proposed a network based on nnUNet and used data augmentation strategies, three types of lesions can be trained separately using the same network. Our results had been validated on the test set of DRAC2022. The main contributions of our work are summarized as follows:

1. Data augmentation was used for raw data to make the number of the three cases close to 1:1:1.
2. To capture more micro lesions, the nnUNet-CBAM based on nnUNet was designed and improved by adding the Convolutional Block Attention Module (CBAM).
3. To solve the problem of overlapping lesions, we trained three segmentation networks to segment three lesions separately.

2   Method

2.1   Preprocess

Before the training phase, a data enhancement strategy was used to IRMAs and NV, random rotation and deformation scaling were used to IRMAs and NV which could make the number of IRMAs and NV is approximately equal to NPAs. In the training phase, we used the same preprocessing pipeline as the heuristic rules generated in nnUNet, including intensity transformation, spatial transformation and data enhancement.

2.2   nnUNet-CBAM Architecture

Our proposed framework is shown in Fig. 2, which is derived from nnUNet. For a segmentation task on a specific dataset, the pipeline of the nnUNet framework mainly consists of the following steps:

1. Collect and analyze data information and generate rule-based parameters using its heuristic rules.
2. Network training based on fixed parameters and rule-based parameters.
3. Integration selection and post-processing methods based on empirical parameters.

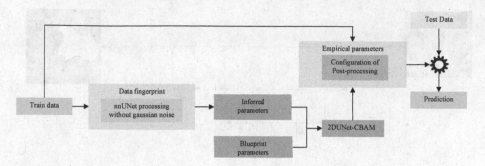

Fig. 2. The nnUNet-CBAM framework: a supervised learning framework based on the nnUNet architecture. The proposed framework follows nnUNet, and replaces the base network from 2DUNet to 2DUNet-CBAM.

2.3 Network Architecture

Our network is an encoder-decoder architecture similar to UNet [9], which has a backbone based on residual blocks [5]. Considering the input data, only 2DUNet is used. 2DUNet can only take one slice as input and needs to be resized to 256*256 during the training and inference phase. The network consists of four encoder modules, four decoder modules and skipping connections, each module contains two convolutional layers, two dropout layers, two BN layers, two LeakyRelu layers, and no pooling layers are used. As illustrated in Fig. 3, the image undergoes a 3×3 convolution. The CBAM module is added each encoder and decoder.

2.4 CBAM Module

Convolutional Block Attention Module (CBAM) is a lightweight attention module that can perform attention operations in both spatial and channel dimensions, CBAM consists of two main parts, channel attention (CA) and spatial attention (SA) modules.

As shown in Fig. 4, the input feature $F \in R^{C*H*W}$, followed by the one-dimensional convolution $M_c \in R^{C*1*1}$ of the channel attention module, multiplying the convolution result by the original map, and using the CAM output result as input, the two-dimensional convolution $M_s \in R^{1*H*W}$ of the spatial attention module, and then multiplying the output result with the original map. The formula is as follows:

$$F' = M_c(F) \ominus F \tag{1}$$

$$F'' = M_s(F') \ominus F' \tag{2}$$

where C is the number of channels, H*W is the size of the input features, F is defined as input feature maps of the channel attention block and $M_c(F)$ is the output of the channel attention block, F' is assigned as input feature maps of the spatial attention block and the $M_s(F')$ is the output of the spatial attention block.

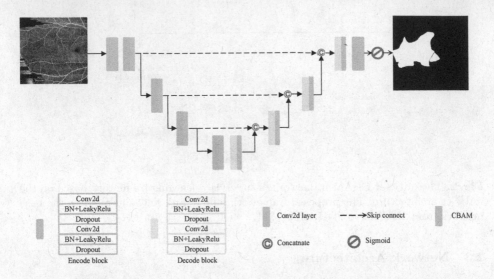

Fig. 3. An encoder-decoder architecture based on 2DUNet generated by the heuristic rules of nnUNet. The network consists of encoder blocks (containing LeakyReLU), decoder blocks (containing inverse ReLU and transposed convolution) and CBAM blocks and the output block.

3   Experiment

3.1   Dataset and Evaluation Measures

The DRAC2022 Challenge provides a standardized ultra-wide (swept source) optical coherence tomography angiography (UW-OCTA) dataset. The dataset includes images and labels for automated image quality assessment, lesion segmentation, and DR grading. We selected the lesion segmentation dataset for

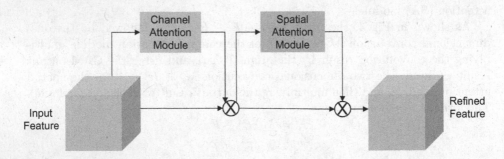

Fig. 4. Convolutional Block Attention Module.

training. We use all images for training, No validation set, including 109 original images with 86 intraretinal microvascular abnormalities, 106 non-perfused areas, and 35 neovascularization cases. We submitte the training results to the test set for evaluation. The evaluation metrics are mean dice similarity coefficient (mDSC) and mean intersection over union (mIOU).

3.2 Implementation Details

In this paper, we choose PyTorch [8] to implement our model and use an NVIDIA GeForce P40*2 GPU for training. The input size of the networks is 256*256 with a batch size of 8. In our model, we set the epoch to 200, and the strategy of five-fold cross-validation was used, using cross-entropy loss and dice loss as the loss function during training, and the optimizer used Adam and set the learning rate dynamic adjustment strategy.

4 Results

To verify the useful of the CBAM module could improve the segmentation results of the lesions, an ablation experiment was designed, as shown in Table 1.

Table 1. DSC score and IOU score of the proposed method and the baseline methods on the test set.

Method	IOU			DSC		
	IRMAs	NPAs	NV	IRMAs	NPAs	NV
nnUNet	0.2202	0.5050	0.3328	0.3264	0.6445	0.4520
nnUNet-CBAM	0.2462	0.5258	0.3359	0.3659	0.6677	0.4554

Table 1 shows the segmentation results of different categories of diabetic retinopathy. We can conclude that the value of mDSC is 0.4963 and the value of mIOU is 0.3693. It is obvious that the segmentation results of IRMAs are not ideal, we could find that IRMAs has the lowest scores in DSC and IOU. As illustrated in Fig. 5, we can find that the lesion of IRMAs has a more severe foreground-background imbalance problem, while the lesion of NPAs has blocky features and obtains the best segmentation results. Finally, the NV lesions are also blocky, but the area of NV lesions is smaller than that of NPAs, and the segmentation result of NV is lower than NPAs but higher than IRMAs.

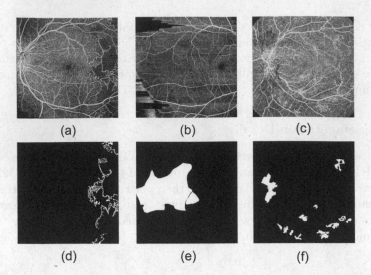

Fig. 5. Three types of DR lesions. (a) Original image of IRMAs. (b) Original image of NPAs. (c) Original image of NV. (d) Groundtruths of IRMAs. (e) Groundtruths of NPAs. (f) Groundtruths of NV.

5   Conclusion

In this work, we proposed a nnUNet-CBAM for lesion segmentation. We constructed UNet-CBAM to extract more details and adopt a supervised learning framework based on nnUNet, in order to address the imbalance of the dataset, we applied data enhancement strategies. Furthermore, we designed an ablation experiment to verify the effect of CBAM. We evaluated our method on the DRAC2022 dataset to demonstrate the effectiveness of the framework. Future work aims to apply our method to other medical image segmentation scenarios.

References

1. Ciulla, T.A., Amador, A.G., Zinman, B.: Diabetic retinopathy and diabetic macular edema: pathophysiology, screening, and novel therapies. Diabetes Care **26**(9), 2653–2664 (2003)
2. L, Dai, et al.: A deep learning system for detecting diabetic retinopathy across the disease spectrum. Nat. Commun. **12**(1), 1–11 (2021)
3. Gulshan, V., et al.: Development and validation of a deep learning algorithm for detection of diabetic retinopathy in retinal fundus photographs. JAMA **316**(22), 2402–2410 (2016)
4. Gulshan, V., et al.: Performance of a deep-learning algorithm vs manual grading for detecting diabetic retinopathy in India. JAMA Ophthalmol. **137**(9), 987–993 (2019)
5. He, K., Zhang, X., Ren, S., Sun, J.: Deep residual learning for image recognition. In: Proceedings of the IEEE Conference on Computer Vision and Pattern Recognition, pp. 770–778 (2016)

6. Kou, C., Li, W., Liang, W., Yu, Z., Hao, J.: Microaneurysms segmentation with a u-net based on recurrent residual convolutional neural network. J. Med. Imaging **6**(2), 025008 (2019)
7. Liu, R., et al.: Deepdrid: Diabetic retinopathy-grading and image quality estimation challenge. Patterns. 100512 (2022)
8. Paszke, A., et al.: PyTorch: an imperative style, high-performance deep learning library. In: Advances in Neural Information Processing Systems, vol. 32 (2019)
9. Ronneberger, O., Fischer, P., Brox, T.: U-Net: convolutional networks for biomedical image segmentation. In: Navab, N., Hornegger, J., Wells, W.M., Frangi, A.F. (eds.) MICCAI 2015. LNCS, vol. 9351, pp. 234–241. Springer, Cham (2015). https://doi.org/10.1007/978-3-319-24574-4_28
10. Russell, J.F., et al.: Longitudinal wide-field swept-source oct angiography of neovascularization in proliferative diabetic retinopathy after panretinal photocoagulation. Ophthalmol. Retina **3**(4), 350–361 (2019)
11. Sánchez, C.I., Hornero, R., Mayo, A., García, M.: Mixture model-based clustering and logistic regression for automatic detection of microaneurysms in retinal images. In: Medical Imaging 2009: Computer-Aided Diagnosis, vol. 7260, pp. 479–486. SPIE (2009)
12. Schaal, K.B., Munk, M.R., Wyssmueller, I., Berger, L.E., Zinkernagel, M.S., Wolf, S.: Vascular abnormalities in diabetic retinopathy assessed with swept-source optical coherence tomography angiography widefield imaging. Retina **39**(1), 79–87 (2019)
13. Sheng, B., et al.: An overview of artificial intelligence in diabetic retinopathy and other ocular diseases. Front. Public Health. **10** (2022)
14. Stanga, P.E., et al.: New findings in diabetic maculopathy and proliferative disease by swept-source optical coherence tomography angiography. OCT Angiogr. Retinal Macul. Dis. **56**, 113–121 (2016)
15. Tian, M., Wolf, S., Munk, M.R., Schaal, K.B.: Evaluation of different Swept'source optical coherence tomography angiography (SS-octa) slabs for the detection of features of diabetic retinopathy. Acta Ophthalmol. **98**(4), e416–e420 (2020)
16. Zhang, Q., Rezaei, K.A., Saraf, S.S., Chu, Z., Wang, F., Wang, R.K.: Ultra-wide optical coherence tomography angiography in diabetic retinopathy. Quant. Imaging Med. Surg. **8**(8), 743 (2018)

An Ensemble Method to Automatically Grade Diabetic Retinopathy with Optical Coherence Tomography Angiography Images

Yuhan Zheng[1]([⊠])(iD), Fuping Wu[2](iD), and Bartłomiej W. Papież[2]([⊠])(iD)

[1] Department of Engineering, University of Oxford, Oxford, UK
yuhan.zheng@spc.ox.ac.uk
[2] Big Data Institute, University of Oxford, Oxford, UK
fuping.wu@ndph.ox.ac.uk,bartlomiej.papiez@bdi.ox.ac.uk

Abstract. Diabetic retinopathy (DR) is a complication of diabetes, and one of the major causes of vision impairment in the global population. As the early-stage manifestation of DR is usually very mild and hard to detect, an accurate diagnosis via eye-screening is clinically important to prevent vision loss at later stages. In this work, we propose an ensemble method to automatically grade DR using ultra-wide optical coherence tomography angiography (UW-OCTA) images available from Diabetic Retinopathy Analysis Challenge (DRAC) 2022. First, we adopt the state-of-the-art classification networks, i.e., ResNet, DenseNet, EfficientNet, and VGG, and train them to grade UW-OCTA images with different splits of the available dataset. Ultimately, we obtain 25 models, of which, the top 16 models are selected and ensembled to generate the final predictions. During the training process, we also investigate the multi-task learning strategy, and add an auxiliary classification task, the Image Quality Assessment, to improve the model performance. Our final ensemble model achieved a quadratic weighted kappa (QWK) of 0.9346 and an Area Under Curve (AUC) of 0.9766 on the internal testing dataset, and the QWK of 0.839 and the AUC of 0.8978 on the DRAC challenge testing dataset.

Keywords: Diabetic retinopathy · Model ensemble · Multi-tasking · Optical Coherence Tomography Angiography

1 Introduction

Diabetic retinopathy (DR) is the most common complication of diabetes and remains a leading cause of visual loss in working-age populations [1]. The current diagnosis pathway relies on the early detection of microvascular lesions [2], such as microaneurysms, hemorrhages, hard exudates, non-perfusion and neovascularization. Based on its clinical manifestations, DR can be divided into two

B. Sheng and M. Aubreville (Eds.): MIDOG 2022/DRAC 2022, LNCS 13597, pp. 46–58, 2023.
https://doi.org/10.1007/978-3-031-33658-4_6

stages, i.e., non-proliferative DR (NPDR) and proliferative DR (PDR), corresponding to early and more advanced stages of DR, respectively. NPDR tends to have no symptoms, and it could take several years [3] to deteriorate into PDR, leading to severe vision impairment. Therefore, detection of early manifestation of DR is essential to accurate diagnosis and treatment monitoring.

Optical coherence tomography angiography (OCTA), as a non-invasive imaging technique providing depth-resolved images for retinal vascular structure [4], has become an effective imaging modality for DR diagnosis. However, manually identifying subtle changes on eye images is difficult and time-consuming. Hence, computer-aided diagnosis of DR using OCTA images has attracted interest from researchers, and many computational approaches have been proposed in recent years, including traditional machine learning approaches such as decision tree and support vector machine [5–7], and convolutional neural networks (CNNs) [8–14]. Previous works can be mainly divided into two categories: (1) segmenting different types of lesions related to DR, and (2) classifying/ grading DR. However, identifying the best performing method requires the standardized datasets. Therefore, to foster the development of image analysis and machine learning techniques in clinical diagnosis of DR, and to address the lack of standardized datasets to make a fair comparison between the developed methods, the Diabetic Retinopathy Analysis Challenge (DRAC) is organized in conjunction with the Medical Image Computing and Computer Assisted Intervention Society (MICCAI) 2022.

In this work, we propose a deep learning model to automatically grade DR from ultra-wide (UW)-OCTA images into three classes, i.e., Normal, NPDR and PDR, with the aim to reduce the burden on ophthalmologists as well as providing a more robust tool to diagnose DR. We investigate four state-of-the-art CNNs for classification with different augmentation techniques, and an ensemble method is employed to generate final DR grading. We also utilize multi-task learning with the help of an auxiliary task i.e. Image Quality Assessment during the training process to further boost the performances of the main task, i.e., DR grading.

The remainder of this paper is organized as follows. Section 2.1 introduces the dataset and the preprocessing steps. The basic models used for DR grading, the ensemble strategy, and multi-task techniques are described in Sect. 2.2, 2.3, and 2.4, respectively. Section 3 describes the experimental set-up and the results for ablation and evaluation. Finally, the discussion and conclusion can be found in Sect. 4. The implementation of our method has been released via our GitHub repository.[1]

2 Methods

2.1 Dataset

DRAC2022 includes three tasks (and the associated dataset) as follows. Task 1: Segmentation Task provides 109 training images containing three types of lesions

---

[1] https://github.com/Yuhanhani/DR-Grading-DRIMGA-.git.

- intraretinal microvascular abnormalities, nonperfusion areas and neovascularization, and 65 testing images. Task 2: Image Quality Assessment contains 665 training images of three classes - Poor, Good and Excellent quality, and 438 testing images. Task 3: Diabetic Retinopathy Grading contains 611 images that is a subset of the previous 665 images (Task 2), and they are divided into three classes - Normal, NPDR and PDR. For challenge testing, 386 testing images are provided for DR grading task, however no expert annotations were available to the participants.

All of the images are grayscale images with the size of 1024 × 1024 pixels. The instrument used in this challenge is a swept-source (SS)-OCTA system (VG200D, SVision Imaging, Ltd., Luoyang, Henan, China), works near 1050 nm nm and features a combination of industry-leading specifications including ultrafast scan speed of 200,000 AScans per second [15].

Our models were mainly developed on DR grading dataset (Task 3). To develop and optimize models, we split the 611 training images into three subsets. Figure 1 illustrates the way they were split, where 84%, 8%, 8% form training, internal validation, and internal testing dataset respectively. The internal validation dataset was used to tune hyperparameters, select optimized models as well as for early stopping to prevent over-fitting. The testing dataset was used as a final sanity check before uploading the models to the challenge competition as well as for comparison between different networks, augmentations, and ensemble strategies presented below. It is also worth mentioning that the overall class distributions of the 611 images are highly imbalanced, with the three classes occupying 53.8% for Normal, 34.7% for NPDR, and 11.5% for PDR, respectively. Therefore, to generate our internal validation and testing dataset, we performed a stratified sampling [16]. Each of the three subsets was guaranteed to follow the same distribution as the overall distribution, so the internal validation and testing dataset can better represent the entire population that was being studied.

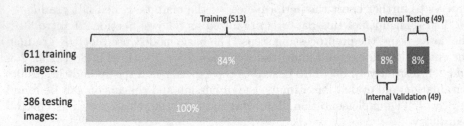

Fig. 1. The splitting strategy on the released training dataset. All of the 611 images were split into three sets: 513 images for training, 49 images for internal validation and 49 images for internal testing. For testing, 386 testing images are provided for DR grading task, but with no expert annotations available to the participants.

2.2   Baseline Networks

We adopted four state-of-the-art CNN architectures, including ResNet [17], DenseNet [18], EfficientNet [19] and VGG [20].

ResNet. This is one of the most commonly used networks that achieves high performance on classification tasks. Many deep learning methods for retinal images classification are developed using ResNet as a backbone [21–23]. ResNet is a deep residual learning framework, which addresses vanishing-gradient and degradation problem, and eases the training of deeper networks by using residual mapping [17].

DenseNet. DenseNet was the another network used for our DR grading, as it is reported to have better feature use efficiency with fewer parameters than ResNet. It utilizes dense blocks, in which each layer is connected to every other layer in a feed-forward manner. It also alleviates the problem of vanishing-gradient, and increases feature reuse and reduces the number of parameters required [18].

EfficientNet. EfficientNet is a set of models obtained by uniformly scaling all dimensions of depth, width, and resolution using compound coefficients. EfficientNet is shown to achieve better performances on ImageNet, and also to transfer well on other datasets [19]. It was also employed on our dataset and brought the benefits of small model size with fast computation speed.

VGG. We finally tried VGG, which is an early CNN architecture that forms the basis of object recognition models. By utilizing small convolutional filters and increasing the depth of CNNs to 16–19 weight layers, it shows outstanding performances on ImageNet [20]. It is also used on OCTA imaging showing good performances [24]. However, its downsides are the large model size and long computation time.

2.3   Model Ensemble

Following the splitting manner introduced in Sect. 2.1, we additionally split the entire dataset randomly three times (A, B, C), such that each time we had different training and internal validation set, while the internal testing set was kept unchanged. Then for each split, all models were trained, and optimized to select the best performing models. The Table 1 summarizes the selection of 16 models used for ensemble method.

Model ensemble is a machine learning technique to combine outputs produced by multiple models in the prediction process, and it overcomes drawbacks associated with a single estimator such as high variance, noise, and bias [25]. In this work, three ensemble strategies were adopted. The first one was plurality voting, which takes the class that receives the highest number of votes as the

Table 1. Summary of all the selected models for given data split (A, B, and C). The number of models selected for each architecture resulting in 16 models in total used for ensemble method.

split	ResNet34	ResNet152	DenseNet121	VGG19_BN	EfficientNet_b0
A	one model	one model	None	one model	one model
B	None	one model	two models	two models	two models
C	two models	None	one model	one model	one model

final prediction [26]. The second was averaging, which outputs the final probabilities as the unweighted average of probabilities estimated by each model. The third technique was label fusion with a three-layer simple neural network, which was trained with inputs as the predictions from the 16 models, and learns to assign appropriate weights to each individual model. An example illustrating these three strategies using five models and three classes can be found in Fig. 2. It can be seen that the final prediction output can be completely different depending on ensemble strategy used. Figure 3 shows our ensemble method for DR grading.

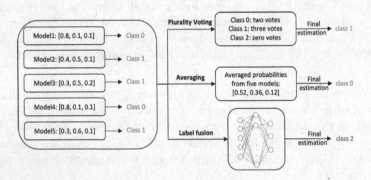

Fig. 2. Diagram shows the example of each ensemble strategy using five models for three-class classification. The predicted probabilities for each class are shown in square brackets.

2.4 Multi-task Learning

We used multi-task learning throughout our network training process, in order to take advantage of complementary information from different tasks, and to improve the robustness of the model. We simultaneously trained model for Image Quality Assessment and Diabetic Retinopathy Grading via hard parameter sharing [27], as shown in Fig. 4. Particularly, Image Quality Assessment Task was treated as an auxiliary task, with the aim to obtain a more generalized model

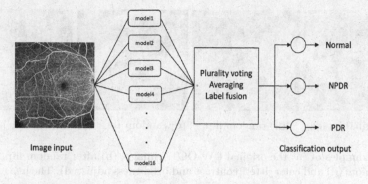

Fig. 3. Schematic diagram of model ensemble strategies where an image is input and the final DR grading is generated by aggregating 16 models.

thereby improving model performances for the main task, i.e., DR grading. The loss function was defined as follows.

$$Loss_{total} = Loss_{task3} + \lambda \cdot Loss_{task2} \qquad (1)$$

where λ is a hyperparameter. In our experiment, $\lambda = 0, 0.01, 0.1, 1$ was the set of values that were tested to improve the performances of DR grading task in the different splits (A,B,C).

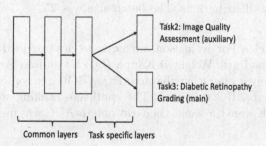

Fig. 4. Schematic diagram of proposed multi-task learning.

3 Experiments

3.1 Experimental Setting

Image Preprocessing. For all experiments, the images were resized to either 256×256 or 512×512 pixels from its original size of 1024×1024 pixels. We further employed image augmentation techniques including random horizontal flip, random vertical flip, random rotation, and color jitter. Figure 5 shows examples of UW-OCTA images with augmentation applied.

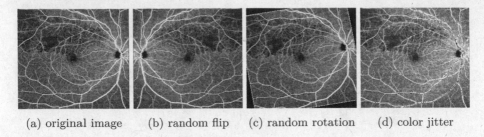

(a) original image   (b) random flip   (c) random rotation   (d) color jitter

Fig. 5. Examples of (a) the original UW-OCTA image, (b) after random flip, (c) random rotation, (d) and color jitter (contrast and brightness adjusted). The images shown here were resized to the size of 256 × 256 pixels.

Training Hyper-parameters All the models were pretrained on ImageNet Dataset [28]. Throughout the training process, we used the stochastic gradient descent (SGD) optimizer with a learning rate scheduler with exponential decay. The initial learning rate was set to 0.001 and the decay factor was set to 0.8 or 0.9. We trained the network using 20 epochs and saved the model that gave the best performance measured by a quadratic weighted kappa (QWK). For $Loss_{task3}$, a weighted cross-entropy loss was used. We increased penalty mostly for NPDR class, as it was the most difficult class to predict for most of the cases. The weights were adjusted empirically. For $Loss_{task2}$, the weighted cross entropy loss was used with the weight values of 0.779, 0.146, and 0.075, based on the Task 2 class distributions. The batch size was 25.

Evaluation Metrics For evaluation, we used metrics provided by the challenge organizer i.e. a Quadratic Weighted Kappa (QWK) and an Area Under Curve (AUC) of the receiver operating characteristic (ROC) curve. Specifically, One-vs-One (ovo) macro-AUC was used. The challenge ranking was based on the QWK, and if QWK were the same, then ovo-macro-AUC was used as a secondary ranking metric.

3.2   Evaluation of Image Preprocessing

Here we study how data augmentation i.e. color jitter and resizing influence the model performance. As shown in Table 2a, for all of the networks except DenseNet, the use of color jitter boosts the performance. Therefore, we incorporated color jitter into our data augmentation process. In addition, Table 2b shows that resizing images down to the size of 256 × 256 pixels did not lead to noticeable performance degradation compared to those with the original image size. The use of lower image resolution made it possible to train with deeper networks and larger batch size. In contrast, with the size of 128 × 128 pixels, the performance of the tested models was consistently lower, which could be due to the loss of information related to the disease.

Table 2. Quadratic Weighted Kappa (QWK) for different models (a) with and without color jitter. (b) and with the varying size of the image, evaluated on the internal validation dataset (average across all splits).

(a) results of applying color jitter

	with	without
ResNet 34	**0.7492**	0.7310
DenseNet 121	0.7659	**0.7850**
VGG19_BN	**0.7969**	0.7932
EfficientNet_b0	**0.7707**	0.7695

(b) results of resizing image

Size	128^2	256^2	512^2	1024^2
ResNet 34	0.5315	0.7492	**0.7713**	0.7561
DenseNet 121	0.5922	0.7659	0.7965	**0.8569**

3.3 Evaluation of Model Ensemble

Here we present the results of different ensemble methods. Table 3a and 3b show the performances of the individual networks and the ensemble methods, evaluated on the internal testing dataset. For the individual model evaluation, VGG19_BN obtained the best performances (the QWK of 0.8715 and the AUC of 0.9665) comparing to the other models. For the ensemble method evaluation, label fusion produced the best result (the QWK of 0.9346 and the AUC of 0.9766) out of the three ensemble techniques. The use of the ensemble method boosted the result by about 6% when compared to single VGG19_BN.

Table 3. The Quadratic Weighted Kappa (QWK) and the Area Under Curve (AUC) of the individual networks used in this study, and three ensemble models, evaluated on the internal testing dataset (average across all splits).

(a) Individual networks

Networks	QWK	AUC
ResNet 34	0.8663	0.9423
ResNet 152	0.8170	0.9350
DenseNet 121	0.8487	0.9583
VGG19_BN	**0.8715**	**0.9665**
EfficientNet_b0	0.8270	0.9420

(b) Ensemble models

Ensemble method	QWK	AUC
Plurality voting	0.9131	0.9766
Averaging	0.9131	0.9677
Label fusion	**0.9346**	**0.9766**

3.4 Evaluation of Multi-task Learning

Table 4 shows the effect of multi-tasking of the three different ensemble strategies. It can be seen that single task outperformed multi-tasking in two cases when using our internal testing dataset. However, it will be shown in Table 5 in

Sect. 3.5 that multi-tasking boosted the performance by 2% on the challenge testing dataset. The reason for this discrepancy could be the fact that our internal testing dataset had only 49 OCTA images, while there were 386 challenge testing images. Therefore, an increase of 2% on the larger challenge testing dataset was considered far more significant to conclude that the multi-tasking can improve the overall performances.

Table 4. The Quadratic Weighted Kappa (QWK) and the Area Under Curve (AUC) of single task and multi-tasking using the three ensemble methods, evaluated on the internal testing dataset.

	Strategies	QWK	AUC		Strategies	QWK	AUC
Single Task	Plurality voting	0.9167	0.9710	Multi-tasking	Plurality voting	0.9131	0.9766
	Averaging	0.9351	0.9677		Averaging	0.9131	0.9677
	Label fusion	0.9322	0.9677		Label fusion	0.9346	0.9766

3.5 Final Challenge Evaluation

The results of our four final submissions to the DRAC challenge testing dataset are shown in Table 5. It can be seen that the best QWK of 0.8390 and the AUC of 0.8978 were achieved using averaging with multi-tasking loss function. It should be noticed that the presence of multi-task learning boosted the final testing result by about 2%. In addition, for ensemble methods, averaging produced the best result as opposed to the internal testing case where label fusion was the best.

Table 5. The Quadratic Weighted Kappa (QWK) and the Area Under Curve (AUC) for different training and ensemble strategies on the DRAC challenge testing dataset.

	Strategies	QWK	AUC		Strategies	QWK	AUC
Single Task	Plurality voting	0.8160	0.9046	Multi-tasking	Plurality voting	0.8308	0.8998
	Averaging	0.8204	0.9062		**Averaging**	**0.8390**	**0.8978**
	Label fusion	0.8089	0.8928		Label fusion	0.8276	0.8922

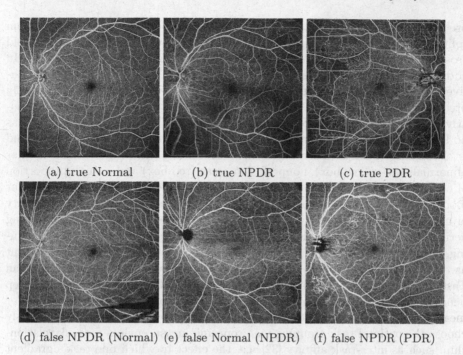

(a) true Normal   (b) true NPDR   (c) true PDR

(d) false NPDR (Normal) (e) false Normal (NPDR)   (f) false NPDR (PDR)

Fig. 6. Examples of six images with labels predicted by our final ensemble model. Image (a),(b) and (c) are three correctly predicted images belonging to Normal, NPDR, PDR respectively. Image (d), (e) and (f) are three incorrectly predicted images with their true labels shown in the bracket. The colored boxes in Image (b) and (c) highlight different types of lesions with yellow, red and blue box indicating intraretinal microvascular abnormalities, nonperfusion areas and neovascularization respectively. They are drawn based on the ground truth provided for the Segmentation Task. (Color figure online)

4   Discussion and Conclusion

In this work, we presented the results of our investigation for the best method to perform diabetic retinopathy grading automatically. We used the state-of-the-art classification networks i.e. ResNet, DenseNet, EfficientNet and VGG as our baseline networks. We selected 16 models with different data splitting strategies, and utilized the ensemble method to produce the final prediction. We showed that hyperparameter tuning, data augmentation, and the use of the multi-task learning, with the auxiliary task of Image Quality Assessment, boosted the main task performances. Our final submission to the DRAC challenge produced the QWK of 0.8390 and the AUC of 0.8978.

By comparing Table 3 and Table 4 it can be observed that the results on our internal testing dataset were always higher than on the challenge testing dataset. This might be due to the fact that, there were only 49 testing images in our internal dataset, which could present distribution bias against the final testing dataset, and our models might have been over-fitted. Therefore, in the future, pretraining on other public OCTA datasets could be used. The three random splits we used, also may also produce optimistic results, hence a nested k-fold cross validation should be attempted to give more generalized results [29]. In addition, attention mechanism could be employed to pick up the most informative region [30]. Furthermore, Fig. 6 shows three correctly and three incorrectly predicted images by the final model. The three correctly predicted images are relatively standard Normal, NPDR and PDR images. Figure 6a is Normal and has no lesions. Figure 6b has some nonperfusion areas (red box) near the corner and is classified as NPDR. There are some surrounding intraretinal microvascular abnormalities (yellow box), nonperfusion areas (red box) and a small proportion of neovascularization (blue box) on Fig. 6c, which conclude being PDR case. Our model, however, predicted incorrectly for the cases such as poorer image quality or non-obvious lesion appearances. This issue could be alleviated by e.g. adopting active learning strategy, which determines hard samples to be used for retraining. Additionally, to regularize the model we chose Image Quality Assessment as the auxiliary task, which is intuitively less correlated to the main task than Task 1, i.e., Segmentation of Diabetic Retinopathy Lesions. However, for deep neural networks, inter-task relationships [31] could be difficult to be defined or measured. Recently, many researches have been done in this area to rigorously choose the optimal task grouping via proposing measurements for task relationship, such as inter-task affinity [32], i.e. the effect to which one task's gradient would affect another task's loss. Hence, we could study in the future how the three challenge tasks could benefit each other by exploring their relationships in multi-task learning framework. Generally, our main task benefited from implicit data augmentation and regularization [33] of multi-tasking, which made it more robust against random noises, and managed to learn more general features, thus yielding a better overall performance.

Acknowledgment. The authors acknowledge the DRAC2022 challenge for available UW-OCTA images for this study [34]. The authors would like to thank Dr. Le Zhang from University of Oxford for helpful comments on our manuscript.

References

1. Wang, W., Lo, ACY.: Diabetic Retinopathy: Pathophysiology and Treatments. Int. J. Mol. Sci. **19**(6), 1816 (2018). https://doi.org/10.3390/ijms19061816
2. Mookiah, M.R.K., Acharya, U.R.A., Chua, C.K., Lim, C.M., Ng, E.Y.K., Laude, A.: Computer-aided diagnosis of diabetic retinopathy: a review. Comput. Biol. Med. **43**(12), 2136–2155 (2013). https://doi.org/10.1016/j.compbiomed.2013.10.007
3. Overview, Diabetic Retinopathy. https://www.nhs.uk/conditions/diabetic-retinopathy/
4. Khalili Pour, E., Rezaee, K., Azimi, H., et al.: Automated machine learning-based classification of proliferative and non-proliferative diabetic retinopathy using optical coherence tomography angiography vascular density maps. Graefes Arch. Clin. Exp. Ophthalmol. (2022). https://doi.org/10.1007/s00417-022-05818-z

5. Abdelsalam, M., Zahran, M.A.: A Novel approach of diabetic retinopathy early detection based on multifractal geometry analysis for OCTA macular images using support vector machine, pp. 22844–22858. IEEE Access (2021). https://doi.org/10.1109/ACCESS.2021.3054743

6. Selvathi, D., Suganya, K.: Support vector machine based method for automatic detection of diabetic eye disease using thermal images. In: 2019 1st International Conference on Innovations in Information and Communication Technology (ICIICT), pp. 1–6 (2019). https://doi.org/10.1109/ICIICT1.2019.8741450

7. Alzami, F., et al.: Diabetic retinopathy grade classification based on fractal analysis and random forest. In: 2019 International Seminar on Application for Technology of Information and Communication (iSemantic), pp. 272–276 (2019). https://doi.org/10.1109/ISEMANTIC.2019.8884217

8. Eladawi, N., et al.: Early signs detection of diabetic retinopathy using optical coherence tomography angiography scans based on 3D multi-path convolutional neural network. In: 2019 IEEE International Conference on Image Processing (ICIP), pp. 1390–1394 (2019). https://doi.org/10.1109/ICIP.2019.8803031

9. Abdelmaksoud, E., et al.: Automatic diabetic retinopathy grading system based on detecting multiple retinal lesions. IEEE Access. **9**, 15939–15960 (2021). https://doi.org/10.1109/ACCESS.2021.3052870

10. Heisler, M. et al.: Ensemble deep learning for diabetic retinopathy detection using optical coherence tomography angiography. Transl. Vision Sci. Technol. **9**(20), 15939–15960 (2021). https://doi.org/10.1167/tvst.9.2.20

11. Ryu, G., et al.: A deep learning model for identifying diabetic retinopathy using optical coherence tomography angiography. Sci. Rep. **11**, 1–9 (2021). https://doi.org/10.1038/s41598-021-02479-6

12. Dai, L., Wu, L., Li, H., et al.: A deep learning system for detecting diabetic retinopathy across the disease spectrum. Nat. Commun. **12**(1), 1–11 (2021). https://doi.org/10.1038/s41467-021-23458-5

13. Liu, R., Wang, X., Wu, Q., et al.: DeepDRiD: diabetic retinopathy-grading and image quality estimation challenge. Patterns **3**(6), 100512 (2022). https://doi.org/10.1016/j.patter.2022.100512

14. Sheng, B., Chen, X., Li, T., Ma, T., Yang, Y., Bi, L., Zhang, X.: An overview of artificial intelligence in diabetic retinopathy and other ocular diseases. Front. Public Health (2022). https://doi.org/10.3389/fpubh.2022.971943

15. SVision. https://svisionimaging.com/index.php/en_us/home/

16. Sechidis, K., Tsoumakas, G., Vlahavas, I.: On the stratification of multi-label data. In: Gunopulos, D., Hofmann, T., Malerba, D., Vazirgiannis, M. (eds.) ECML PKDD 2011. LNCS (LNAI), vol. 6913, pp. 145–158. Springer, Heidelberg (2011). https://doi.org/10.1007/978-3-642-23808-6_10

17. He, K., Zhang, X., Ren, S., Sun, J.: Deep Residual Learning for Image Recognition. CoRR (2015). arXiv:1512.03385

18. Huang, G., Liu, Z., Maaten, L., Weinberger, K.: Densely Connected Convolutional Networks. CoRR(2016). arXiv:1608.06993

19. Tan, M., Le, Q.: EfficientNet: Rethinking Model Scaling for Convolutional Neural Networks. CoRR(2019). arXiv:1905.11946

20. Simonyan, K., Zisserman, A: Very deep convolutional networks for large-scale image recognition. In: ICLR (2015). arxiv:1409.1556

21. Dhodapkar, R.M., et al.: Deep learning for quality assessment of optical coherence tomography angiography images. Sci. Rep. (2022). https://doi.org/10.1038/s41598-022-17709-8

22. Jin, K., et al.: Multimodal deep learning with feature level fusion for identification of choroidal neovascularization activity in age-related macular degeneration. Acta Ophthalmologica. **100**(2), 512–520 (2021). https://doi.org/10.1111/aos.14928

23. Padmasini, N. et al.: Automated detection of multiple structural changes of diabetic macular oedema in SDOCT retinal images through transfer learning in CNNs. IET Image Process. **14**(16), 4067–4075 (2021). https://doi.org/10.1049/iet-ipr.2020.0612

24. Le, D., et al.: Transfer learning for automated OCTA detection of diabetic retinopathy. Transl. Vision Sci. Technol. **9**(2) (2020). https://doi.org/10.1167/tvst.9.2.35

25. Pham, K.: Ensemble learning-based classification models for slope stability analysis. CATENA **196** (2021). https://doi.org/10.1016/j.catena.2020.104886

26. Mu, X., Watta, P., Hassoun, M.: Analysis of a plurality voting-based combination of classifiers. Neural Process. Lett. **29**, 89–107 (2009)

27. Ruder, S.: An Overview of Multi-Task Learning in Deep Neural Networks. CoRR (2017). arXiv:1706.05098

28. Deng, J., et al.: ImageNet: a large-scale hierarchical image database. In: 2009 IEEE Conference on Computer Vision and Pattern Recognition, pp. 248–255 (2009). CVPR.2009.5206848

29. Cawley, G.C., Talbot, N.L.C.: On over-fitting in model selection and subsequent selection bias in performance evaluation. J. Mach. Learn. Res. **11**, 2079–2107 (2010)

30. Bourigault, E., et al.: Multimodal PET/CT tumour segmentation and prediction of progression-free survival using a full-scale UNet with attention. In: Andrearczyk, V., Oreiller, V., Hatt, M., Depeursinge, A. (eds.) Head and Neck Tumor Segmentation and Outcome Prediction. HECKTOR 2021. LNCS, vol. 13209, pp. 189–201. Springer, Cham (2021). https://doi.org/10.1007/978-3-030-98253-9_18

31. Liu, S., et al.: Auto-Lambda: Disentangling Dynamic Task Relationships. CoRR (2022). arxiv:2202.03091

32. Fifty, C., et al.: Efficiently Identifying Task Groupings for Multi-Task Learning. CoRR (2021). arxiv:2109.04617

33. Ruder, S.: An Overview of Multi-Task Learning in Deep Neural Networks. CoRR (2017). arxiv:1706.05098

34. Sheng, B., et al.: Diabetic retinopathy analysis challenge 2022. In: 25th International Conference on Medical Image Computing and Computer Assisted Intervention (MICCAI 2022) (2022). https://doi.org/10.5281/zenodo.6362349

Bag of Tricks for Developing Diabetic Retinopathy Analysis Framework to Overcome Data Scarcity

Gitaek Kwon, Eunjin Kim, Sunho Kim, Seongwon Bak, Minsung Kim, and Jaeyoung Kim[✉]

VUNO Inc., Seoul, South Korea
{gitaek.kwon,eunjin.kim,ksunho0660,seongwon.bak,
minsung.kim,jaeyoung.kim}@vuno.co

Abstract. Recently, diabetic retinopathy (DR) screening utilizing ultra-wide optical coherence tomography angiography (UW-OCTA) has been used in clinical practices to detect signs of early DR. However, developing a deep learning-based DR analysis system using UW-OCTA images is not trivial due to the difficulty of data collection and the absence of public datasets. By realistic constraints, a model trained on small datasets may obtain sub-par performance. Therefore, to help ophthalmologists be less confused about models' incorrect decisions, the models should be robust even in data scarcity settings. To address the above practical challenging, we present a comprehensive empirical study for DR analysis tasks, including lesion segmentation, image quality assessment, and DR grading. For each task, we introduce a robust training scheme by leveraging ensemble learning, data augmentation, and semi-supervised learning. Furthermore, we propose reliable pseudo labeling that excludes uncertain pseudo-labels based on the model's confidence scores to reduce the negative effect of noisy pseudo-labels. By exploiting the proposed approaches, we achieved 1st place in the Diabetic Retinopathy Analysis Challenge (Code is available at https://github.com/vuno/DRAC22_MICCAI_FAI).

Keywords: Diabetic Retinopathy Analysis · Semi-supervised learning

1 Introduction

Diabetic retinopathy (DR) is an eye disease that can result in vision loss and blindness in people with diabetes, but early DR might cause no symptoms or only mild vision problems [7]. Therefore, early detection and management of DR play a crucial role in improving the clinical outcome of eye condition. Color fundus photography, fluorescein angiography (FA), and optical coherence tomography angiography (OCTA) have been used in diabetic eye screening to acquire valuable information for DR diagnosis and treatment planning. Recently, in the screening, ultra-wide OCTA (UW-OCTA) images have been widely used leveraging their advantages such as more detailed visualization of vessel structures,

B. Sheng and M. Aubreville (Eds.): MIDOG 2022/DRAC 2022, LNCS 13597, pp. 59–73, 2023.
https://doi.org/10.1007/978-3-031-33658-4_7

and ability to capture a much wider view of the retinal compared to previous standard approaches [36].

With the advancements of deep learning (DL), applying DL-based methods for medical image analysis has become an active research area in the ophthalmology fields [13, 24, 29, 30]. Notably, the availability to large amounts of annotated fundus photography has been one of the key elements driving the quick growth and success of developing automated DR analysis tools. Sun et al. [31] develop the automatic DR diagnostic models using color fundus images, and Zhou et al. [38] propose a collaborative learning approach to improve the accuracy of DR grading and lesion segmentation by semi-supervised learning on the color fundus photography. Although previous studies investigate the effectiveness of applying DL to DR grading and lesion detection tasks based on color fundus images, DR analysis tool leveraging UW-OCTA are still under-consideration. One of the reasons lies in the fact that annotating high-quality UW-OCTA images is inherently difficult because the annotation of medical images requires manual labeling by experts. Consequently, when we consider about the practical restrictions, it is one of the most crucial things to develop a robust model even in the lack of data.

To address the above real-world setting, we introduce the bag of tricks for DR analysis tasks using the Diabetic Retinopathy Analysis Challenge (DRAC22) dataset, which consists of three tasks (i.e., lesion segmentation, image quality assessment, and DR grading) [27]. To alleviate the negative effect introduced by the lack of labeled data, we investigate the effectiveness of data augmentations, ensembles of deep neural networks, and semi-supervised learning. Furthermore, we propose reliable pseudo labeling (RPL) that selects reliable pseudo-labels based on a trained classifier's confidence scores, and then the classifier is re-trained with labeled and trustworthy pseudo-labeled data.

In our study, we find that Deep Ensembles [11], test-time data augmentation (TTA), and RPL have powerful effects for DR analysis tasks. Our solutions are combinations of the above techniques and achieved 1st place in all tasks for DRAC22.

2   Related Work

In this section, we overview previous studies on the DR analysis (Sect. 2.1), and semi-supervised learning algorithms (Sect. 2.2).

2.1   Diabetic Retinopathy Analysis

Automatic DR assessment methods based on neural networks have been developed to assist ophthalmologists [4, 15, 21, 22, 26, 37]. Gulshan et al. [8] develop the convolution neural network (CNN) for detecting DR, and the proposed method shows the competitive result with ophthalmologists in detection performance. They demonstrates the feasibility of the DL-based computer-aided diagnosis system for fundus photography. Dai et al. [4] suggest a unified framework called

DeepDR in order to improve the interpretability of CNNs. DeepDR provides comprehensive predictions, including DR grade, location of DR-related lesions, and an image quality assessment of color fundus photography.

On the other hand, a series of approaches based on the FA [5, 18], OCT [6, 10], and OCTA [23, 35] have been studied to detect DR. Pan et al. [18] propose the CNN-based model, which classifies DR findings (i.e., non-perfusion regions, microaneurysms, and laser scars) with FA. Heisler et al. [10] suggest an ensemble network for DR classification. Each ensemble member is trained with OCT and OCTA, respectively. For a testing time, they use aggregated predictions of the ensemble model to provide robust and calibrated predictions.

Although the previous methods have shown remarkable results in promoting the accuracy of DR grading, a comprehensive empirical study of applying UW-OCTA to DL has yet to be conducted.

2.2 Semi-supervised Learning

In the medical imaging domain, collecting labeled data is challenging due to expensive costs and time-consuming. Instead, it is much easier to obtain unlabeled data. Thanks to the recent success of semi-supervised learning (SSL), various SSL algorithms [2, 19] show impressive performance on various tasks such as semantic segmentation, object detection, and image recognition.

Pseudo-labeling (PL) [12] is a simple and effective method in SSL approaches, in which pseudo labels are generated based on the pretrained-network's predictions, and then the network is re-trained both labeled and pseudo-labeled data simultaneously. Following the pioneering approach of pseudo-labeling, Sohn et al. [28] propose FixMatch, which produces pseudo labels using both consistency regularization and pseudo-labeling. FixMatch only retains a pseudo label if the network produces a high probability for a weak-augmented image in order to reduce an error of the prediction caused by the distortions of a given image. Xie et al. [34] suggests an iterative training scheme for SSL, called noisy student training. In their process, they first train a model on labeled data and use it as a teacher network to generate pseudo labels for unlabeled data. They then train an equal-or-larger model as a student network on the combination of labeled and pseudo-labeled samples and iterate the above process by assigning the student as the teacher.

3   Method

In this section, we first introduce a simple and effective technique, reliable pseudo labeling (RPL), for improving classification performance in the data scarcity setting. Then, we describe our solutions for each task in detail.

Scope of the Paper. We consider a standard multi-class classification problem which comprise a common setting in computer vision tasks. In this paper, our goal is for a model to classify as $\hat{y}_i = \{1, ..., C\} \in \mathbb{N}$ (i.e., discrete random variable) for a given x_i.

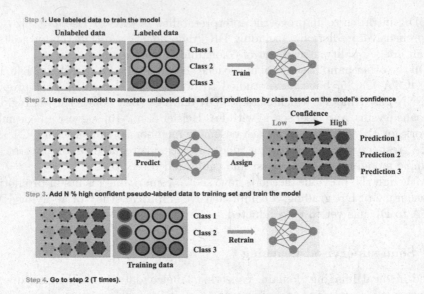

Step 1. Use labeled data to train the model

Fig. 1. Overall procedure of RPL.

3.1 Reliable Pseudo Labeling

Pitfall of Pseudo-labels. While PL shows powerful performance in the data scarcity setting, networks can produce an incorrect prediction on unseen data. If the model is trained using incorrect pseudo-labels, errors accumulate and confirmation bias can appear since modern over-parameterized neural networks easily overfit to noisy samples [1]. Hence, it would be reckless to consider all pseudo-labels generated by a network trained with a small amount of data as correct predictions. To address the vulnerability of PL, we propose RPL, and overall procedure is described in Fig. 1.

Notation. Let $[m] := \{1, ..., m\}$, $\sigma(\cdot)$ is the softmax function, and $\lfloor \cdot \rceil$ is the nearest integer function. We denote a given dataset by $\mathcal{D} = \mathcal{D}^l \cup \mathcal{D}^u$, where $\mathcal{D}^l = \{(x_i^l, y_i^l) : i \in [n]\}$ is the labeled set and $\mathcal{D}^u = \{x_i^u : i \in [m]\}$ is the unlabeled set. For multi-class classification tasks, a softmax classifier f_{cls} maps an input $x_i \in \mathbb{R}^{W \times H}$ into a predictive distribution $\hat{p}(y|\sigma(z_i))$, where z_i is a vector of logits $f_{\text{cls}}(x_i)$ and $y \in [C]$ is a discrete class label. When the classification task is formulated by regression problem, a class prediction of a regressor f_{reg} can be calculated by $\lfloor f_{\text{reg}}(x_i) \rceil$ when a label is defined as $y \in \mathbb{N}$.

Procedure of RPL. We first train f using an arbitrary loss function using $\mathcal{D}^l_{\text{train}}$. After training, we collect pseudo-labeled set per predicted class $\hat{y}$ defined as $\hat{\mathcal{D}}^u = \{\hat{\mathcal{D}}^u_k\}_{k=1}^C$, where $\hat{\mathcal{D}}^u_k = \{(x_i^u, \hat{y}_i^u)|\hat{y}_i^u = k, i \in [m]\}$. Then we sort each $\hat{\mathcal{D}}_k$ by the model's confidence score corresponding the predicted class. For the softmax classifier, the confidence score can be $\max_c \hat{p}(y = c|\sigma(z_i))$. When the classification task is formulated as regression, the confidence score can be calculated with $||\lfloor f_{\text{reg}}(x_i) \rceil - f_{\text{reg}}(x_i)|^1$.

To exclude uncertain data, we select N_c^t (%) confident pseudo labeled samples in each $\hat{\mathcal{D}}_c^u$, and then re-train the model which learns both labeled and trustworthy pseudo-labeled data. Finally, we repeat the process T times. For each process, N_c^t is increased by $N_c^t = 100\frac{t \cdot |\mathcal{D}_c^u|}{T}$, where $t = \{1, ..., T\}$ is an indicator of the process.

3.2 Overview of Solutions

For readers' convenience, we provide a brief description of our solutions in Table 1. In the rest of this section, we introduce our solutions in detail.

Table 1. Summary of methods used in experiments. Unless otherwise specified, we use reported methods and their parameters for all experiments. MPA [16]: Multi-scale patch aggregation.

	Task 1	Task 2	Task 3
Preprocessing	Dividing all pixels by 255		
Input resolution	1024 × 1024		
Post-processing	IRMA: false positive removal NP: Dilation (kernel=5) NV: false positive removal	Class-specific thresholding	Post-editing using the segmentation model
Test-time augmentation	IRMA: rotate=$\{90, 180, 270, 360\}$. NP: rotate=$\{90, 180, 270, 360\}$. NV: rotate=$\{90, 180, 270, 360\}$, $MPA = \{1.0, 1.1, 1.2, 1.3, 1.4\}$	Flip=$\{$horizontal, vertical$\}$	Flip=$\{$horizontal, vertical$\}$
Architecture	U^2-Net	EfficientNet-b2	EfficientNet-b2
Pretrained weights	-	ImageNet	ImageNet
Loss	Weighted dice loss, focal loss, binary cross entropy loss	Smooth L1	Smooth L1
Optimizer	AdamW [17]	AdamW	AdamW
Learning rate	1e-4 (w/o scheduler)	2e-4 (w/o scheduler)	2e-4 (w/o scheduler)
Augmentation	see Table 2	see Table 3	see Table 4
Weight decay	1e-2	1e-2	1e-2
SSL	-	RPL (T=5)	RPL (T=5)
Dropout ratio	0.0	0.2	0.2
Batch size	2	8	8
Epochs	400	150	150
Train/Dev split	1:0	0.8:0.2	0.8:0.2
Ensemble	NP: 5 models, IRMA and NV: w/o ensemble	5 models	5 models

---

[1] In this paper, we only consider the classification task with a *discrete* label space

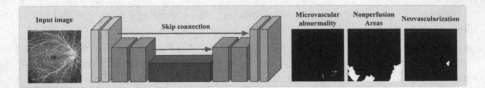

Fig. 2. For lesion segmentation task, we train the U-shape network f_{seg}. For training, normalized input image x_i is fed into the network, and f_{seg} are trained with empirical risk minimization.

3.3   Lesion Segmentation (Task 1)

Motivation. The goal of DR-related lesion segmentation task in DRAC22 is detecting the pixel-level lesions, including intraretinal microvascular abnormalities (IRMA), nonperfusion areas (NP), and neovascularization (NV). Although one unified segmentation model formulated by multi-label classification can detect the locations of three lesions, the model's detection performance may be sub-optimal since the anatomical characteristics are different among the lesions. For example, IRMA and NV often appear as small objects, and such imbalanced data distribution may sensitively affects the data-driven deep learning methods [33]. On the other hand, several studies [9] find that a segmentation model confuses NP and signal reduction artifacts. It is natural to think that hard example mining may reduce the false positive for the regions of signal artifacts. Hence, we design models focusing on imbalanced data setting for small lesions (i.e., IRMA and NV) and hard example mining for NP, respectively (Fig. 2).

Training. We train two independent U^2-Net models [20], one is small-lesion segmentation network $f_{\text{seg}}^{\text{small}}$ for IRMA and NV, another is NP segmentation model $f_{\text{seg}}^{\text{NP}}$. Each model learn to minimize the difference between the predicted lesion maps and the ground-truth masks:

$$\mathcal{L}_{\text{total}}(y_i, \sigma(f(x_i))) = \mathcal{L}_{\text{Dice}}(y_i, \sigma(f(x_i))) + \alpha \mathcal{L}_{\text{Aux}}(y_i, \sigma(f(x_i))), \quad (1)$$

where $\sigma(\cdot)$ is the sigmoid function, $\mathcal{L}_{\text{Dice}}$ is the weighted dice loss, $\mathcal{L}_{\text{Aux}}$ is the auxiliary loss, and α is the hyper-parameter that determines the magnitude of the auxiliary loss. $\mathcal{L}_{\text{Dice}}$ is calculated by:

$$\mathcal{L}_{\text{Dice}}(y, \hat{y}) = 1 - 2 \times \frac{\sum_{c=1}^{C=3} w_c \sum_i y_i^c \cdot \hat{y}_i^c}{\sum_{c=1}^{C=3} w_c \sum_i (y_i^c + \hat{y}_i^c)}, \quad (2)$$

where $\hat{y} = \sigma(f(x))$ is the prediction given x, C is the number of classes, $w_c = \log \frac{1}{\sum_i y_i^c}$, $i = \{1, ..., W \times H\}$ is the class-wise weight.

We use focal loss [14] as $\mathcal{L}_{\text{Aux}}$ for $f_{\text{seg}}^{\text{NP}}$ and modify the original focal loss in order to apply multi-label classification:

$$\mathcal{L}_{\text{Focal}}(y_i^c, \hat{y}_i^c) = \begin{cases} -(1 - \hat{y}_i^c) \log \hat{y}_i^c, & \text{if } y^c = 1, \\ -\hat{y}_i^c \log(1 - \hat{y}_i^c), & \text{otherwise.} \end{cases} \tag{3}$$

For $f_{\text{seg}}^{\text{small}}$, we set binary cross-entropy loss as $\mathcal{L}_{\text{Aux}}$ which less penalizes false positive pixels compared to Eq. 3. In summary, $f_{\text{seg}}^{\text{NP}}$ is trained with strong penalties not only false positive instances but also hard-to-distinguish pixels, whereas $f_{\text{seg}}^{\text{small}}$ is trained focused on positive instances. Training configurations are summarized in Table 1. We set α as 0.5 for all experiments.

Table 2. The list of augmentations for training lesion segmentation models. Each transformation can be easily implemented by albumentation library [3]. The input image has 1024×1024 resolution.

Operator	Parameters	Probability
RandomBrightnessContrast (ω_1)	brightness_limit=0.2, contrast_limit=0.2	1.0
RandomGamma (ω_2)	gamma_limit=(80, 120)	
Sharpen (ψ_1)	alpha=(0.2, 0.5), lightness=(0.5, 1.0)	1.0
Blur (ψ_2)	blur_limit=3	
Downscale (ψ_3)	scale_min=0.7, scale_max=0.9	
Flip (ϕ_1)	horizontal, vertical	0.5
ShiftScaleRotate (ϕ_2)	shift_limit=0.2, scale_limit=0.1, rotate_limit=90	0.5
GridDistortion (ϕ_3)	num_steps=5, distort_limit=0.3	0.2
CoarseDropout (ϕ_4)	max_height=128, min_height=32, max_width=128, min_width=32, max_holes=3	0.2

Augmentation. For all segmentation models, we use the following data augmentation strategy. Let $\mathcal{A} = \mathcal{A}_{\text{pixel}} \cup \mathcal{A}^c$ be a set of n augmentations. $\mathcal{A}_{\text{pixel}} = \{\Omega, \Psi\}$ is the subset of $\mathcal{A}$. Ω and Ψ respectively have child operators, i.e., $\Omega = \{\omega_k\}_{k=1}^m$, and $\Psi = \{\psi_k\}_{k=1}^{m'}$. For training the segmentation model, we always apply pixel-wise transformations, and an augmented image is defined as $\bar{x}_i = \Psi^*(\Omega^*(x_i))$, where Ω^* and Ψ^* are randomly picked from Ω and Ψ, respectively.

To generate diverse input representations, we also apply geometric transform ϕ_k to $\bar{x}_i$, and ϕ_k is randomly sampled from $\mathcal{A}^c = \{\phi_k\}_{k=1}^{n-(m+m')}$. As a result, segmentation models are trained with $\mathcal{D}_{\text{seg}} = \{(\phi(\bar{x}_i), y_i)\}_{i=1}^{|\mathcal{D}_{\text{seg}}|}$, thus, each model never encounter original training samples. The list of operators and detailed parameters are described in Table 2.

Ensemble. To boost the performance, we use ensemble techniques such as TTA and Deep Ensemble [11].

- IRMA: TTA-averaging the predictions of $f_{\text{seg}}^{\text{small}}$ across multiple rotated samples of data-is used.
- NP: We use an averaged prediction of five independent models' prediction where each prediction also applied TTA with rotation operators.

– NV: TTA with rotation transformation and MPA [16] are used.

Post-processing. To reduce the incorrect prediction, the following post-processing methods are used. We denote the prediction masks as $\mathbf{P}_{IRMA}$, $\mathbf{P}_{NP}$, and $\mathbf{P}_{NV}$, respectively.

– NP: A prediction mask is applied dilation operation with a kernel size of 5.
– IRMA: To reduce the false positive pixels, we replace a positive pixel $p_{ij} \in \mathbf{P}_{IRMA}$ to negative where $q_{ij} \in \mathbf{P}_{NV}$ have more confident prediction compared to p_{ij}.
– NV: In the same way as above, a positive pixel q_{ij} is replaced with zero when p_{ij} is more confident at the same region.

3.4   Image Quality Assessment (Task 2) and DR Grading (Task 3)

Motivation. The goal of image quality assessment and DR grading task in DRAC22 is to distinguish qualities of UW-OCTA images and grading the severity of DR. Image quality consists of three levels: poor quality level (PQL), good quality level (GQL), and Excellent quality level (EQL), and the DR grade consists of three levels: normal, non-proliferatived diabetic retinopathy (NPDR), and proliferatived diabetic (PDR). We formulate the above tasks as a regression problem rather than a multi-class classification problem in order to consider the correlation among classes.

Training. For each task, we build the regression model f_{reg} using the EfficientNet-b2 [32] initialized with pretrained weights for ImageNet. To address the lack of labeled training data, RPL is applied. We use final test set in DRAC22 as unlabeled dataset. The detailed hyper-parameters and training configurations are reported in Table 1.

Table 3. The list of augmentations for task 2.

Operator	Parameters	Probability
RandomBrightnessContrast (ω_1)	brightness_limit=0.2, contrast_limit=0.2	**1.0**
RandomGamma (ω_2)	gamma_limit=(80, 120)	
Sharpen (ψ_1)	alpha=(0.2, 0.5), lightness=(0.5, 1.0)	**1.0**
Blur (ψ_2)	blur_limit=3	
Downscale (ψ_3)	scale_min=0.7, scale_max=0.9	
Flip (ϕ_1)	horizontal, vertical	0.5
ShiftScaleRotate (ϕ_2)	shift_limit=0.2, scale_limit=0.1, rotate_limit=45	0.5

Augmentation. The strategy of augmentation is the same as in Sect. 3.3, but task 2 and task 3 have different combinations of operators, respectively. The detailed components are reported in Table 3, and Table 4.

Ensemble. For testing-time, we use an averaged prediction of five independent models' prediction. Also each prediction of the model is generated by TTA with flip operators.

Post-processing. The following post-processing methods are used.

Table 4. The list of augmentations for task 3.

Operator	Parameters	Probability
RandomBrightnessContrast (ω_1)	brightness_limit=0.2, contrast_limit=0.2	1.0
RandomGamma (ω_2)	gamma_limit=(80, 120)	
Sharpen (ψ_1)	alpha=(0.2, 0.5), lightness=(0.5, 1.0)	1.0
Blur (ψ_2)	blur_limit=3	
Downscale (ψ_3)	scale_min=0.7, scale_max=0.9	
Flip (ϕ_1)	horizontal, vertical	0.5
ShiftScaleRotate (ϕ_2)	shift_limit=0.2, scale_limit=0.1, rotate_limit=45	0.5
CoarseDropout (ϕ_3)	max_height=5, min_height=1, max_width=512, min_width=51, max_holes=5	0.2

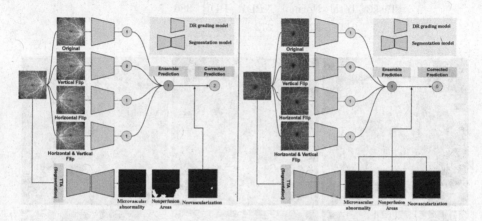

Fig. 3. Summarization of the post-processing process of task 3. The lesion segmentation model is used to correct incorrect prediction labels.

- Task 2: We use the following decision rule using class-specific operating thresholds instead of $\hat{y} = \lfloor f_{\mathrm{reg}}(x) \rceil$:

$$\hat{y} = \begin{cases} 0, & \text{if } f_{reg}(x) < 0.54, \\ 1, & \text{if } 0.54 \leq f_{reg}(x) < 1.5, \\ 2, & \text{otherwise.} \end{cases} \quad (4)$$

- Task 3: Retrospectively, we find that the DR grading model ignores the NV region (NV is a sure sign of PDR) when the region is small in PDR samples. In this case, the model ultimately misclassifies PDR as NPDR, thus, we replace the failure prediction for NPDR to PDR using the segmentation model's prediction mask of NV (left in Fig. 3). In contrast, if the segmentation model predicts normal for all classes, we correct the the DR grading model's prediction to normal (right in Fig. 3).

Table 5. Data statistics for DRAC22. **PQL**: Poor Quality Level, **GQL**: Good Quality Level, **EQL**: Excellent Quality Level, **NPDR**: Non-Proliferatived DR, **PDR**: Proliferatived DR.

	# train				# test (unlabeled)
Task1	Total	IRMA	NP	NV	65
	109	86	106	35	
Task2	Total	PQL	GQL	EQL	438
	665	50	97	518	
Task3	Total	Normal	NPDR	PDR	386
	611	329	212	70	

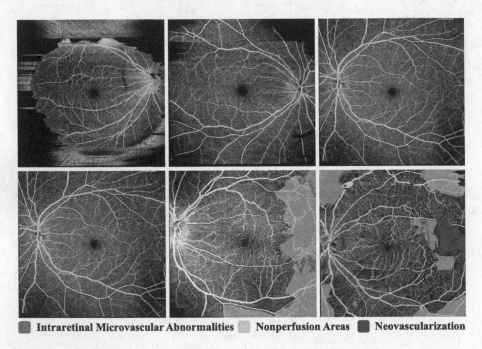

■ **Intraretinal Microvascular Abnormalities** ■ **Nonperfusion Areas** ■ **Neovascularization**

Fig. 4. Visualized example of DRAC22 dataset. (**Top**) Each column respectively represents PQL, GQL, and EQL with respect to task 2. (**Bottom**) In order from left to right, each image represents Normal, NPDR, and PDR, respectively.

4   Experiments

4.1   Dataset and Metrics

Dataset. In DRAC22, the dataset consists of three tasks, i.e., lesion segmentation, image quality assessment, and DR grading[2]. Data statistics are described

---

[2] https://drac22.grand-challenge.org/.

in Table 5, and we also provide an example of DRAC22 dataset (see Fig. 4). We use 20% data as a validation set for task 2 and task 3, and we select the best models when validation performance is highest. Partially, for task 1, we use all training samples so that we select the model which has the highest dice score with respect to the training set.

Metrics. For task 1, the averaged dice similarity coefficient (mean-DSC) and the averaged intersection of union (mean-IoU) are measured to evaluate the segmentation models. For task 2 and task 3, the quadratic weighted kappa (QWK) and Area Under Curve (AUC) are used to evaluate the performance of the proposed methods.

4.2   Results

Table 6. The ablation study results for task 1 on testset of DRAC22. Note that ensemble is only used for f_{seg}^{NP}. Post: Post-processing.

Ensemble	TTA	Post	mean-DSC	mean-IOU	IRMA DSC	NP DSC	NV DSC
			0.5859	0.4311	0.4596	0.6803	0.6179
✓			0.5865	0.4380	0.4596	0.6821	0.6179
✓	✓		0.5927	0.4418	0.4607	0.6911	0.6263
✓	✓	✓	**0.6067**	**0.4590**	**0.4704**	**0.6926**	**0.6571**

Task 1. We report the lesion segmentation performance in Table 6. Our best segmentation model achieves the mean-DSC of 0.6067, the mean-IOU of 0.4590, respectively. Notably, combining ensemble techniques and post-processing indeed shows the effectiveness for the lesion segmentation task.

Table 7. The ablation study results for task 2 on testset of DRAC22.

Ensemble	PL	RPL	TTA	Post	QWK	AUC
					0.7321	0.7487
✓					0.7485	0.7640
✓	✓				0.7757	0.7786
✓		✓			0.7884	0.7942
✓		✓	✓		0.7920	0.7923
✓		✓	✓	✓	**0.8090**	**0.8238**

Task 2. Table 7 presents an ablation study to assess each component of the solution by removing parts as appropriate. In Table 7, the first row is the performance of baseline, which obtain the QWK of 0.7332 and AUC of 0.8492,

Table 8. The ablation study results for task 3 on testset of DRAC22.

Ensemble	PL	RPL	TTA	Post	QWK	AUC
					0.8272	0.8749
✓					0.8557	0.8856
✓	✓				0.8343	0.8630
✓		✓			0.8607	0.8844
✓		✓	✓		0.8684	0.8865
✓		✓	✓	✓	**0.8910**	**0.9147**

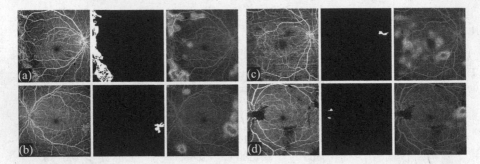

Fig. 5. Localization result of DR grading model for PDR samples including NV. Each column in the block represents original image, ground truth mask for NV, and activation map, respectively. To obtain the activation map, we use GradCAM [25]. (**a-b**) The prediction result of the model is PDR. (**c-d**) The prediction result of the model is NPDR (failure cases).

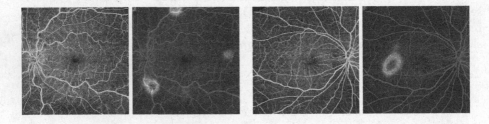

Fig. 6. Failure cases where the model's prediction is NPDR for normal samples.

respectively. Other components are incrementally applied, where performance is enhanced consistently with the addition of each step. Specially, RPL approaches show the noticeable performance improvement with respect to QWK compared to PL.

Task 3. An ablation study is performed with the results shown in Table 8 to evaluate the performance for multiple techniques. The baseline (first row in Table 8) obtains sub-par performance compared to other tricks. Similar to task 2, model ensemble, RPL, TTA, and the post-editing show advanced performance for QWK and AUC, and applying the post-processing is most effective for DR grading.

To analyze why the post-processing is effective, we analyze what areas the DR grading model is focusing on making predictions. Figure 5 shows PDR samples in which the segmentation model detects NV. In the left block of Fig. 5, the DR grading model activates the NV and classifies it as PDR when the NV area is large. On the other hand, when the NV area is small, the DR grading model does not activate the NV and classifies it as NPDR (right in Fig. 5). Therefore, it seems useful to correct the results of the DR grading model to PDR when the segmentation model detects NV. In contrast, as shown in Fig 6, there are cases where the DR grading model activates the artifact area for normal samples in which the segmentation model detects no lesions. If the segmentation model does not detect any DR lesions, it seems reasonable to correct the results of the DR grading model to normal.

5   Conclusion

In this paper, we present a fully automated DR analysis system using UW-OCTA. We find that various tricks including ensemble learning, RPL, and TTA show advanced performance compared to baselines, and RPL shows significantly improved performance in a data scarcity setting compared to a naive pseudo-labeling. By assembling these tricks, we achieved 1st place in the DRAC22. We hope our study can revitalize the field of UW-OCTA research and the proposed methods serve as a strong benchmark for DR analysis tasks. In addition, we expect our approaches to substantially benefit clinical practices by improving the efficiency of diagnosing DR and reducing the workload of DR screening.

References

1. Arazo, E., Ortego, D., Albert, P., O'Connor, N.E., McGuinness, K.: Pseudo-labeling and confirmation bias in deep semi-supervised learning. In: 2020 International Joint Conference on Neural Networks (IJCNN), pp. 1–8, IEEE (2020)
2. Berthelot, D., et al.: Remixmatch: Semi-supervised learning with distribution alignment and augmentation anchoring. arXiv preprint arXiv:1911.09785 (2019)
3. Buslaev, A., Iglovikov, V.I., Khvedchenya, E., Parinov, A., Druzhinin, M., Kalinin, A.A.: Albumentations: fast and flexible image augmentations. Information. **11**(2) (2020). https://doi.org/10.3390/info11020125, https://www.mdpi.com/2078-2489/11/2/125
4. Dai, L., et al.: A deep learning system for detecting diabetic retinopathy across the disease spectrum. Nat. Commun. **12**(1), 1–11 (2021)
5. Gao, Z., et al.: End-to-end diabetic retinopathy grading based on fundus fluorescein angiography images using deep learning. Graefes Arch. Clin. Exp. Ophthalmol. **260**(5), 1663–1673 (2022)
6. Ghazal, M., Ali, S.S., Mahmoud, A.H., Shalaby, A.M., El-Baz, A.: Accurate detection of non-proliferative diabetic retinopathy in optical coherence tomography images using convolutional neural networks. IEEE Access **8**, 34387–34397 (2020)

7. Gregori, N.Z.: Diabetic retinopathy: Causes, symptoms, treatment. Am. Acad. Ophthalmol. (2021)
8. Gulshan, V., et al.: Development and validation of a deep learning algorithm for detection of diabetic retinopathy in retinal fundus photographs. JAMA **316**(22), 2402–2410 (2016)
9. Guo, Y., Camino, A., Wang, J., Huang, D., Hwang, T.S., Jia, Y.: Mednet, a neural network for automated detection of avascular area in oct angiography. Biomed. Opt. Express **9**(11), 5147–5158 (2018)
10. Heisler, M., et al.: Ensemble deep learning for diabetic retinopathy detection using optical coherence tomography angiography. Transl. Vision Sci. Technol. **9**(2), 20–20 (2020)
11. Lakshminarayanan, B., Pritzel, A., Blundell, C.: Simple and scalable predictive uncertainty estimation using deep ensembles. In: Advances in Neural Information Processing Systems, vol. 30 (2017)
12. Lee, D.H., et al.: Pseudo-label: the simple and efficient semi-supervised learning method for deep neural networks. In: Workshop on Challenges in Representation Learning, ICML, vol. 3, p. 896 (2013)
13. Li, T., et al.: Applications of deep learning in fundus images: a review. Med. Image Anal. **69**, 101971 (2021)
14. Lin, T.Y., Goyal, P., Girshick, R., He, K., Dollár, P.: Focal loss for dense object detection. In: Proceedings of the IEEE International Conference on Computer Vision, pp. 2980–2988 (2017)
15. Liu, R., et al.: Deepdrid: Diabetic retinopathy-grading and image quality estimation challenge. Patterns, p. 100512 (2022)
16. Liu, S., Qi, X., Shi, J., Zhang, H., Jia, J.: Multi-scale patch aggregation (MPA) for simultaneous detection and segmentation. In: Proceedings of the IEEE Conference on Computer Vision and Pattern Recognition, pp. 3141–3149 (2016)
17. Loshchilov, I., Hutter, F.: Decoupled weight decay regularization. arXiv preprint arXiv:1711.05101 (2017)
18. Pan, X., et al.: Multi-label classification of retinal lesions in diabetic retinopathy for automatic analysis of fundus fluorescein angiography based on deep learning. Graefes Arch. Clin. Exp. Ophthalmol. **258**(4), 779–785 (2020)
19. Pham, H., Dai, Z., Xie, Q., Le, Q.V.: Meta pseudo labels. In: Proceedings of the IEEE/CVF Conference on Computer Vision and Pattern Recognition, pp. 11557–11568 (2021)
20. Qin, X., Zhang, Z., Huang, C., Dehghan, M., Zaiane, O.R., Jagersand, M.: U2-net: going deeper with nested u-structure for salient object detection. Pattern Recogn. **106**, 107404 (2020)
21. Qummar, S., et al.: A deep learning ensemble approach for diabetic retinopathy detection. IEEE Access **7**, 150530–150539 (2019)
22. Ruamviboonsuk, P., et al.: Deep learning versus human graders for classifying diabetic retinopathy severity in a nationwide screening program. NPJ Digital Med. **2**(1), 1–9 (2019)
23. Ryu, G., Lee, K., Park, D., Park, S.H., Sagong, M.: A deep learning model for identifying diabetic retinopathy using optical coherence tomography angiography. Sci. Rep. **11**(1), 1–9 (2021)
24. Sarki, R., Ahmed, K., Wang, H., Zhang, Y.: Automatic detection of diabetic eye disease through deep learning using fundus images: a survey. IEEE Access **8**, 151133–151149 (2020)

25. Selvaraju, R.R., Cogswell, M., Das, A., Vedantam, R., Parikh, D., Batra, D.: Grad-cam: Visual explanations from deep networks via gradient-based localization. In: Proceedings of the IEEE International Conference on Computer Vision, pp. 618–626 (2017)
26. Sheng, B., et al.: An overview of artificial intelligence in diabetic retinopathy and other ocular diseases. Front. Public Health. **10** (2022)
27. Sheng, B., et al.: Diabetic retinopathy analysis challenge 2022, March 2022. https://doi.org/10.5281/zenodo.6362349
28. Sohn, K., et al.: Fixmatch: simplifying semi-supervised learning with consistency and confidence. Adv. Neural. Inf. Process. Syst. **33**, 596–608 (2020)
29. Son, J., Park, S.J., Jung, K.H.: Towards accurate segmentation of retinal vessels and the optic disc in fundoscopic images with generative adversarial networks. J. Digit. Imaging **32**(3), 499–512 (2019)
30. Son, J., Shin, J.Y., Kim, H.D., Jung, K.H., Park, K.H., Park, S.J.: Development and validation of deep learning models for screening multiple abnormal findings in retinal fundus images. Ophthalmology **127**(1), 85–94 (2020)
31. Sun, R., Li, Y., Zhang, T., Mao, Z., Wu, F., Zhang, Y.: Lesion-aware transformers for diabetic retinopathy grading. In: Proceedings of the IEEE/CVF Conference on Computer Vision and Pattern Recognition, pp. 10938–10947 (2021)
32. Tan, M., Le, Q.: Efficientnet: rethinking model scaling for convolutional neural networks. In: International Conference on Machine Learning, pp. 6105–6114. PMLR (2019)
33. Xi, X., Meng, X., Qin, Z., Nie, X., Yin, Y., Chen, X.: IA-net: informative attention convolutional neural network for choroidal neovascularization segmentation in oct images. Biomed. Opt. Express **11**(11), 6122–6136 (2020)
34. Xie, Q., Luong, M.T., Hovy, E., Le, Q.V.: Self-training with noisy student improves imagenet classification. In: Proceedings of the IEEE/CVF Conference on Computer Vision and Pattern Recognition, pp. 10687–10698 (2020)
35. Zang, P., et al.: Dcardnet: diabetic retinopathy classification at multiple levels based on structural and angiographic optical coherence tomography. IEEE Trans. Biomed. Eng. **68**(6), 1859–1870 (2020)
36. Zhang, Q., Rezaei, K.A., Saraf, S.S., Chu, Z., Wang, F., Wang, R.K.: Ultra-wide optical coherence tomography angiography in diabetic retinopathy. Quant. Imaging Med. Surg. **8**(8), 743 (2018)
37. Zhang, W., et al.: Automated identification and grading system of diabetic retinopathy using deep neural networks. Knowl. Based Syst. **175**, 12–25 (2019)
38. Zhou, Y., et al.: Collaborative learning of semi-supervised segmentation and classification for medical images. In: Proceedings of the IEEE/CVF Conference on Computer Vision and Pattern Recognition, pp. 2079–2088 (2019)

Deep-OCTA: Ensemble Deep Learning Approaches for Diabetic Retinopathy Analysis on OCTA Images

Junlin Hou[1], Fan Xiao[2], Jilan Xu[1], Yuejie Zhang[1(✉)], Haidong Zou[3], and Rui Feng[1,2(✉)]

[1] School of Computer Science, Shanghai Key Laboratory of Intelligent Information Processing, Fudan University, Shanghai, China
{jlhou18,yjzhang,fengrui}@fudan.edu.cn
[2] Academy for Engineering and Technology, Fudan University, Shanghai, China
21210860085@m.fudan.edu.cn
[3] Department of Ophthalmology, Shanghai General Hospital, School of Medicine, Shanghai Jiao Tong University, Shanghai, China
zouhaidong@sjtu.edu.cn

Abstract. The ultra-wide optical coherence tomography angiography (OCTA) has become an important imaging modality in diabetic retinopathy (DR) diagnosis. However, there are few researches focusing on automatic DR analysis using ultra-wide OCTA. In this paper, we present novel and practical deep-learning solutions based on ultra-wide OCTA for the Diabetic Retinopathy Analysis Challenge (DRAC). In the first task of segmentation of DR lesions, we utilize UNet and UNet++ to segment three lesions with strong data augmentation and model ensemble. In the second task of image quality assessment, we create an ensemble of Inception-V3, SE-ResNeXt, and Vision Transformer models. Pre-training on the large dataset as well as the hybrid MixUp and CutMix strategy are both adopted to boost the generalization ability of our models. In the third task of DR grading, we build a Vision Transformer and find that the model pre-trained on color fundus images serves as a useful substrate for OCTA images. Extensive ablation studies demonstrate the effectiveness of each designed component in our solutions. The proposed methods rank 4th, 3rd, and 5th on the three leaderboards of DRAC, respectively. Our code is publicly available at https://github.com/FDU-VTS/DRAC.

Keywords: Diabetic retinopathy analysis · Optical coherence tomography angiography · Deep learning

1 Introduction

Diabetic Retinopathy (DR) is a chronic progressive disease that causes visual impairment due to retinal microvascular damage. It has become a leading cause

This work was supported (in part) by the Science and Technology Commission of Shanghai Municipality (No. 21511104502).

B. Sheng and M. Aubreville (Eds.): MIDOG 2022/DRAC 2022, LNCS 13597, pp. 74–87, 2023.
https://doi.org/10.1007/978-3-031-33658-4_8

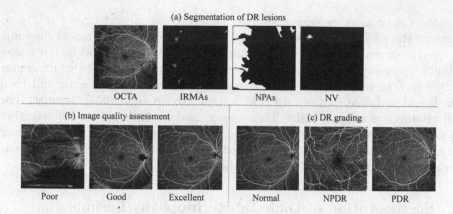

(a) Segmentation of DR lesions

OCTA   IRMAs   NPAs   NV

(b) Image quality assessment   (c) DR grading

Poor   Good   Excellent   Normal   NPDR   PDR

Fig. 1. Examples from the three tasks in the DRAC dataset, including (a) segmentation of DR lesions (b) image quality assessment, and (c) DR grading.

of legal blindness in the working-age population worldwide [20]. DR is diagnosed by the presence of retinal lesions, such as microaneurysms (MAs), intraretinal microvascular abnormalities (IRMAs), nonperfusion areas (NPAs) and neovascularization (NV) [10]. The traditional diagnosis of DR grading mainly relies on fundus photography and fluorescein fundus angiography (FFA). With rising popularity, OCT angiography (OCTA) has become a reliable tool due to its capability of visualizing the retinal and choroidal vasculature at a microvascular level [11]. Compared to fundus photography and FFA, the ultra-wide OCTA can non-invasively detect the changes of DR neovascularization, thus it is an important imaging modality to assist ophthalmologists in diagnosing Proliferative DR (PDR).

Recently, deep learning approaches have achieved promising performance in DR diagnosis [1,5,7,9]. Based on fundus photography, numerous deep learning methods are proposed for lesion segmentation, biomarkers segmentation, disease diagnosis and image synthesis [5]. A deep learning system, named DeepDR, is developed to perform real-time image quality assessment, lesion detection and DR grading [1]. However, there are currently few works capable of automatic DR analysis using ultra-wide OCTA images.

In this paper, we propose novel and practical methods for DR analysis based on ultra-wide OCTA images. Our methods are demonstrated effective in the MICCAI 2022 Diabetic Retinopathy Analysis Challenge (DRAC) [10]. This challenge provides a standardized ultra-wide OCTA dataset, including lesion segmentation, image quality assessment, and DR grading. As illustrated in Fig. 1, the segmentation subset includes three DR lesions, i.e. Intraretinal Microvascular Abnormalities (IRMAs), Nonperfusion Areas (NPAs), and Neovascularization (NV). The image quality assessment subset divides images into three quality levels, i.e. Poor quality level, Good quality level, and Excellent quality level. The DR grading subset contains images of three different DR grades, i.e. Normal, Non-Proliferative DR (NPDR), and Proliferative DR (PDR). In our solutions, we first develop two encoder-decoder networks, namely UNet [8] and UNet++ [21], to segment three lesions separately. Strong data augmentation and model ensemble are demonstrated helpful to the generalization ability of our models. For image quality assessment, we create an ensemble of multiple state-of-the-art neural networks. The networks are first pre-trained on large-scale datasets, and then fine-tuned on DRAC dataset with the hybrid MixUp and CutMix strategy. For DR grading, we adopt a Vision Transformer model, which utilizes self-attention to integrate information across the entire OCTA image. Extensive experiments on DRAC dataset show that our proposed solutions achieve superior DR analysis performance, ranking 4th, 3rd, 5th on the three leaderboards, respectively.

2   Segmentation of Diabetic Retinopathy Lesions

2.1   Methodology

We adopt the UNet [8] and UNet++ [21] networks with pre-trained encoders for DR lesion segmentation. Customized strategies are designed to train the segmentation models of three different lesions. For IRMAs segmentation, Step1 learning rate schedule and color jittering augmentation are both adopted. However, they are not utilized to boost the segmentation performance of NPAs and NV lesions. Besides, we employ the model ensemble strategy when predicting the segmented masks of IRMAs and NPAs. In the following section, we will introduce the UNet and UNet++ networks, learning rate schedules, and loss functions.

Backbone Network. (1) UNet [8] is an encoder-decoder network, where the encoder includes down-sampling layers and the decoder consists of up-sampling layers with skip connections. The feature structures of different levels are combined through skip connections. (2) UNet++ [21] proposes an efficient ensemble of U-Nets of varying depths, which partially share an encoder and co-learn simultaneously using deep supervision. It redesigns skip connections to exploit multi-scale semantic features, leading to a flexible feature aggregation scheme.

Learning Rate Schedule. In order to train our network efficiently, we analyze two different schedules for learning rate decay. One method, named Step1, divides the initial learning rate by 10 at 25% of the total number of training epochs. The other method, named Step2, is to decay the learning rate to $0.6\times$ of the previous value every quarter epoch of the total epochs.

Loss Function. We adopt two loss functions, i.e. Dice loss $\mathcal{L}_D$ and Jaccard loss $\mathcal{L}_J$, to train the segmentation models of DR lesions. Specifically, the Dice loss $\mathcal{L}_D$ is expressed by:

$$\mathcal{L}_D = \frac{1}{N} \sum_{i=1}^{N} \left(1 - \text{Score}_{Dice(i)}\right), \tag{1}$$

Table 1. The results on DRAC dataset of IRMAs segmentation.

ID	Method	Loss	Metric	Schedule	Aug	Val Dice	Test Dice
1	UNet++	Dice	Dice	Step1	All	0.5637	0.4172
2	UNet++	Dice	IoU	Step1	All	0.6145	0.4166
3	UNet++	Jaccard	IoU	Step1	All	0.4648	0.3933
4	UNet++	Dice	Dice	Step2	All	0.6139	0.3788
5	UNet	Dice	Dice	Step2	All	0.5215	0.3514
6	UNet	Dice	Dice	Step2	w/o CJ	0.6115	0.2774
7	Ensemble ID '1,2,4'	-	-	-	-	-	**0.4257**

where N denotes the total number of samples, and $\text{Score}_{Dice(i)}$ is the Dice score of sample x_i. The Jaccard loss $\mathcal{L}_J$ is defined as:

$$\mathcal{L}_J = \frac{1}{N} \sum_{i=1}^{N} \left(1 - \text{Score}_{IoU(i)}\right), \tag{2}$$

where $\text{Score}_{IoU(i)}$ denotes the IoU score of sample x_i. We train the networks by each loss function and a joint loss function for segmentation of three lesions.

2.2 Dataset

DRAC Dataset. The DRAC dataset for segmentation of diabetic retinopathy lesions contains three different diabetic retinopathy lesions, i.e. Intraretinal Microvascular Abnormalities (1), Nonperfusion Areas (2), and Neovascularization (3). Training set consists of 109 images and corresponding lesion masks, where each category includes 86/106/35 images, respectively. We randomly select 20% images of the training set as the validation set. Testing set consists of 65 images and the ground-truth masks are not available during the challenge.

2.3 Implementation Details

All images with the original image size (1024 × 1024) are fed to UNet and UNet++. Strong data augmentation includes horizontal flipping, rotating, random cropping, gaussian noise, perspective, and color jittering. The networks are optimized using the Adam algorithm and trained for 100 epochs. The initial learning rate is set to 1e-4, and we use step learning rate schedule.

2.4 Ablation Study

In this part, we conduct an ablation study to evaluate the effectiveness of each component for segmentation of DR lesions, including backbone network, learning rate schedule, color jittering, metric, and model ensemble.

Table 2. The results on DRAC dataset of NPAs segmentation.

ID	Method	Loss	Metric	Schedule	Aug	Val Dice	Test Dice
1	UNet++	Dice	Dice	Step1	All	0.6211	0.5874
2	UNet++	Dice	IoU	Step1	All	0.6460	0.5942
3	UNet++	Jaccard	IoU	Step1	All	0.6348	0.5930
4	UNet++	Dice	Dice	Step2	All	0.6699	0.6105
5	UNet	Dice	Dice	Step2	All	0.6805	0.6197
6[†]	UNet	Dice	Dice	Step2	All	0.7027	0.6275
7	UNet	Dice&Jaccard	Dice	Step2	All	0.6541	0.6011
8	UNet	Dice	Dice	Step2	w/o CJ	0.6434	0.6261
9	Ensemble ID '4,5,6,7'	-	-	-	-	-	**0.6414**

[†] The network was re-trained by a different random seed.

Table 3. The results on DRAC dataset of NV segmentation.

ID	Method	Loss	Metric	Schedule	Aug	Val Dice	Test Dice
1	UNet++	Dice	Dice	Step1	All	0.2360	0.3838
2	UNet++	Jaccard	IoU	Step1	All	0.1984	0.3741
3	UNet++	Dice	Dice	Step2	All	0.5250	0.4803
4	UNet	Dice	Dice	Step2	All	0.5547	0.5445
5	UNet	Dice	Dice	Step2	w/o CJ	0.4881	**0.5803**

Backbone Network. Both UNet and UNet++ backbone structures are used to segment three DR lesions separately. From the results in the 4th and 5th rows of Table 1, we can see that UNet++ obtains superior IRMAs segmentation performance than UNet by 9.24% val dice and 2.74% test dice. On the contrary, UNet is demonstrated more suitable for the NPAs and NV segmentation. It outperforms UNet++ by a 1.70% improvement for NPAs and a 6.42% improvement for NV on test dice, as can be seen from the 4th and 6th rows of Table 2 and the 3rd and 4th rows of Table 3.

Learning Rate Schedule. We compare the impact of different learning rate schedules for each lesion segmentation. As can be seen in the 1st and 4th rows in Table 1, the Step1 schedule largely outperforms Step2 by 3.84% test dice for IRMAs segmentation. In contrast, the Step1 schedule fails in segmenting the other two lesions. From the 1st and 4th rows of Table 2, we can see that the NPAs segmentation performance obtained by the Step2 schedule greatly surpasses Step1 by 2.31% test dice. For NV segmentation, the results in the 1st and 3rd rows of Table 3 show that the Step2 schedule obtains a significant improvement of 9.65% on test dice.

Color Jittering. We investigate the effectiveness of color jittering (CJ) augmentation. For IRMAs segmentation, applying CJ can greatly improve the performance on test dice by 7.40%, as can be found in the 5th and 6th rows of

Table 4. The top-5 results on the leaderboard of segmentation of DR lesions.

Rank	Team	mDice	mIoU	IRMAs		NPAs		NV	
				Dice	IoU	Dice	IoU	Dice	IoU
1	FAI	0.6067	0.4590	0.4704	0.3189	0.6926	0.5566	0.6571	0.5015
2	KT_Bio_Health	0.6046	0.4573	0.4832	0.3299	0.6736	0.5396	0.6570	0.5024
3	AiLs	0.5756	0.4223	0.4672	0.3144	0.6680	0.5260	0.5917	0.4433
4	FDVTS_DR (ours)	0.5491	0.4041	0.4257	0.2853	0.6414	0.5031	0.5803	0.4240
5	LaTIM	0.5387	0.3966	0.4079	0.2691	0.6515	0.5129	0.5566	0.4077

Table 1. The results demonstrate that CJ is effective to improve the generalization ability of the model. However, for the segmentation of NPAs and NV, we experimentally find that CJ shows no significant improvement on test dice.

Metric. Following the same protocol as the DRAC challenge, we adopt Dice and IoU metrics to evaluate our model on the validation set and choose the best model. As can be observed in the first two rows of all the three tables, the models selected by the two metrics obtain very close segmentation performance for each lesion.

Model Ensemble. To further boost the performance on test samples, we ensemble several models for IRMAs and NPAs segmentation. As shown in Table 1, the ensembled model composed of model ID '1,2,4' (7th row) improves the performance by 0.85% test dice, compared to the best UNet++ (1st row). In the 6th and 9th rows of Table 2, we can observe that the ensembled model of model ID '4,5,6,7' obtains higher test dice than the best UNet by a 1.39% improvement.

2.5 Results on the Leaderboard

Table 4 shows the top-5 results of our method and other participants on the testing set of task 1 in the DRAC challenge. Our ensembled UNet++ for IRMAs, ensembled UNet and UNet++ for NPAs, and single model UNet for NV, with customized learning rate schedules and color jittering augmentation, rank 4th on the leaderboard, approaching 0.5491 mDice and 0.4041 mIoU.

3 Image Quality Assessment

3.1 Methodology

Our proposed model takes an OCTA image $X \in \mathbb{R}^{W \times H \times 3}$ as the input and outputs the image quality prediction $\hat{Y}$ in an end-to-end manner. To alleviate the overfitting problem caused by limited samples, we first pre-train our model on the large-scale OCTA-25K-IQA-SEG dataset. Then, we fine-tune the model on DRAC dataset with the hybrid MixUp and CutMix strategy. We create an ensemble of three models, including Inception-V3, SE-ResNeXt, and Vision Transformer (ViT). In the following section, we will introduce the three models, hybrid MixUp and CutMix strategy, and joint loss function in detail.

Backbone Network. (1) Inception-V3 [13] aims to reduce the parameter count and computational cost by replacing larger convolutions with a sequence of smaller and asymmetric convolution layers. It also adopts auxiliary classifiers with batch normalization as aggressive regularization. (2) SE-ResNeXt [16] is composed of SENet and ResNeXt, where SENet employs the Squeeze-and-Excitation block to recalibrates channel-wise feature responses adaptively, and ResNeXt aggregates a set of transformations with the same topology. (3) ViT [2] receives a sequence of flattened image patches as input into a Transformer encoder, which consists of alternating layers of multi-head attention and MLP blocks.

Hybrid MixUp and CutMix Strategy. We adopt the MixUp strategy [19] to mix training data, where each sample is interpolated with another sample randomly chosen from a mini-batch. Specifically, for each pair of OCTA images (x_i, x_j) and their quality labels (y_i, y_j), the mixed (x', y') is computed by:

$$
\begin{aligned}
\lambda_1 &\sim \text{Beta}(\alpha_1, \alpha_1), \\
x' &= \lambda_1 x_i + (1 - \lambda_1)x_j, \\
y' &= \lambda_1 y_i + (1 - \lambda_1)y_j,
\end{aligned}
\tag{3}
$$

where the combination ratio $\lambda_1 \in [0, 1]$ is sampled from the beta distribution. The hyper-parameter α_1 controls the strength of interpolation of each pairs, and we set $\alpha_1 = 0.4$ in this work.

We also utilize another augmentation strategy named CutMix [17]. It first cuts patches and then pastes them among training images, where the labels are also mixed proportionally to the area of the patches. Formally, let (x_i, y_i) and (x_j, y_j) denote two training samples. The combined (x', y') is generated as:

$$
\begin{aligned}
\lambda_2 &\sim \text{Beta}(\alpha_2, \alpha_2), \\
x' &= \mathbf{M} \odot x_i + (1 - \mathbf{M}) \odot x_j, \\
y' &= \lambda_2 y_i + (1 - \lambda_2)y_j,
\end{aligned}
\tag{4}
$$

where $\mathbf{M} \in \{0,1\}^{W \times H}$ denotes a binary mask indicating where to drop out and fill in, $\mathbf{1}$ is a matrix filled with ones, and $\odot$ is element-wise multiplication. In our experiments, we set α_2 to 1. The mask $\mathbf{M}$ can be determined by the bounding box coordinates $\mathbf{B} = (r_x, r_y, r_w, r_h)$, which are uniformly sampled according to:

$$
\begin{aligned}
r_x &\sim \text{Unif}(0, W), \ r_y \sim \text{Unif}(0, H), \\
r_w &= W\sqrt{1 - \lambda_2}, \ r_h = H\sqrt{1 - \lambda_2}.
\end{aligned}
\tag{5}
$$

With the cropping region, the binary mask $\mathbf{M} \in \{0,1\}^{W \times H}$ is decided by filling with 0 within the bounding box $\mathbf{B}$, otherwise 1.

Joint Loss Function. Different from the original design [17,19] where they replaced the classification loss with the mix loss, we merge the standard cross-entropy classification loss $\mathcal{L}_{clf}$ with the mix loss $\mathcal{L}_{mix}$ to enhance the classification ability on both raw and mixed samples. The joint loss can be calculated as $\mathcal{L} = \mathcal{L}_{clf} + \mathcal{L}_{mix}$. Specifically, the classification loss is expressed by:

$$\mathcal{L}_{clf} = -\frac{1}{N} \sum_{i=1}^{N} y_i^T \log \hat{y}_i, \tag{6}$$

where N denotes the total number of samples, y_i is the one-hot vector of ground truth label, and $\hat{y}_i$ is the predicted probability of sample x_i. For the mix loss, we additionally utilize label smoothing [13] to reduce overfitting:

$$\mathcal{L}_{mix} = -\frac{1}{N} \sum_{i=1}^{N} \tilde{y}_i^{'T} \log \hat{y}_i', \tag{7}$$

where $\hat{y}_i'$ is the predicted probability of mixed sample x_i', and $\tilde{y}_i' = y_i'(1-\epsilon)+\epsilon/K$ represents the smoothed label, K denotes the number of classes. In our work, the hyper-parameter ϵ is set to 0.1.

3.2   Dataset

DRAC Dataset. The DRAC dataset for image quality assessment contains three different levels, i.e. Poor quality level (0), Good quality level (1), and Excellent quality level (2). Training set consists of 665 images and corresponding labels, where each grade includes 50/97/518, respectively. We use 5-fold cross validation. Testing set consists of 438 images and the labels are not available during the challenge.

OCTA-25K-IQA-SEG Dataset. The OCTA-25K-IQA-SEG dataset [15] provides 14,042 $6 \times 6mm^2$ superficial vascular layer OCTA images with annotations of image quality assessment (IQA) and foveal avascular zone (FAZ) segmentation. Similarly, each OCTA image is labeled into three IQA categories, including ungradable (0), gradable (1), or outstanding (2). In this work, we adopt the OCTA-25K-IQA-SEG dataset for network pre-training.

3.3   Implementation Details

All images are resized to 224×224 as the input to SE-ResNeXt, 384×384 for ViT, and 512×512 for other models. Data augmentation includes random cropping, flipping, and color jittering. All images are normalized by mean=0.5 and std=0.5. The networks are optimized using the SGD algorithm and trained for 100 epochs. The initial learning rate is set to 1e-3, and we use cosine annealing learning rate schedule.

Table 5. The results of image quality assessment on DRAC dataset. The val Kappa is the mean value of 5-fold cross validation.

ID	Method	Pre-training	Mix	Val Kappa	Test Kappa
1	Inception-V3 [13]	ImageNet	×	0.8203	0.7817
2	Inception-Res-V2 [12]		×	0.8222	0.6881
3	EfficientNet-B6 [14]		×	0.7866	Unknown
4	ResNeSt-50 [18]		×	0.8450	0.7444
5	ViT-t [2]		×	0.8488	0.7253
6	ViT-s [2]		×	0.8272	0.7298
7	SE-ResNeXt-101 [3]	OCTA-25K-IQA-SEG	×	0.8774	0.7580
8			✓	0.8560	0.7647
9	Inception-v3 [13]		×	0.8576	0.7358
10			✓	0.8189	0.6989
11	ViT-s [2]		×	0.8621	Unknown
12			✓	0.8710	Unknown
13	Ensemble ID '1, 4, 5, 6'	-	-	-	0.7835
14	Ensemble ID '1, 12'	-	-	-	0.7864
15	Ensemble ID '1, 5, 6'	-	-	-	0.7877
16	Ensemble ID '1, 8, 12'	-	-	-	**0.7896**

3.4   Ablation Study

In this part, we conduct an ablation study to understand the influence of each component for image quality assessment, including backbone network, pre-training scheme, hybrid MixUp and CutMix strategy, and model ensemble.

Backbone Network. First, we evaluate the effects of different backbone architectures. As can be seen from the 1st to 6th rows of Table 5, the ResNeSt-50 [18] and ViT-t [2] models achieve superior performance with more than 0.84 Kappa, and the Inception-V3 [13], Inception-Res-V2 [12] and ViT-s [2] models reach over 0.82 Kappa on the 5-fold cross validation. Surprisingly, the Inception-V3 outperforms other models on the testing set, approaching 0.7817 on Kappa score.

Pre-training Scheme. To alleviate the overfitting issue, we first pre-train our models on the large-scale OCTA-25K-IQA-SEG dataset [15]. From the validation results in the 7th, 9th and 11th rows of Table 5, we find the pre-trained model can offer better initialization weights. Compared to ImageNet pre-training, the Inception-V3 and ViT-s models achieve 3.73% and 3.49% improvements on val Kappa, respectively. The SE-ResNeXt-101 [3] reaches the best performance with 0.8774 val Kappa.

Hybrid MixUp and CutMix Strategy. In addition, we investigate the impact of hybrid Mixup and CutMix strategy in the 7th to 12th rows of Table 5. It can be seen that this strategy successfully increases the test Kappa by 0.67%

Table 6. The top-5 results on the leaderboard of image quality assessment.

Rank	Team	AUC	Kappa
1	FAI	0.8238	0.8090
2	KT_Bio_Health	0.9085	0.8075
3	FDVTS_DR (ours)	0.9083	0.7896
4	LaTIM	0.8758	0.7804
5	yxc	0.8874	0.7775

for SE-ResNeXt-101, as well as the val Kappa by 0.89% for ViT-s. However, we experimentally find that the strategy cannot benefit the performance of Inception-V3.

Model Ensemble. Finally, we ensemble several models to further boost the performance of image quality assessment. The final prediction of each OCTA image is obtained by averaging the predictions from individual models. We show the results of different model combinations in the 13th to 16th rows of Table 5. The ensembled model, which is composed of ID '1,8,12' models, surpasses all the other models, approaching 0.7896 Kappa on the testing set.

3.5   Results on the Leaderboard

Table 6 presents the top-5 results of our method and other participants on the testing set of task 2 in the DRAC challenge. Our final method, an ensemble of Inception-V3, SE-ResNeXt-101, and ViT-s, ranked 3rd on the leaderboard with 0.9083 AUC and 0.7896 Kappa.

4   Diabetic Retinopathy Grading

4.1   Methodology

We adopt a Vision Transformer (ViT) model for DR grading, which receives an OCTA image and produces the grading score in an end-to-end manner. Following the same training procedure in image quality assessment, we first pre-train our model on EyePACS [4] and DDR [6] datasets, which provide a large number of color fundus images for DR grading. We then fine-tune the model on DRAC dataset with the hybrid MixUp and CutMix strategy. The model is trained to minimize a joint objective function of a mix loss and a classification loss.

4.2   Dataset

DRAC Dataset. The DRAC dataset for diabetic retinopathy grading contains three grades, i.e. Normal (0), NPDR (1), PDR (2). Training set consists of 611 images and corresponding labels, where each grade includes 329/212/70, respectively. We use 5-fold cross validation. Testing set consists of 386 images and the labels are not available during the challenge.

Table 7. The results of diabetic retinopathy grading on DRAC dataset. The val Kappa is the mean value of 5-fold cross validation.

ID	Method	Pre-training	Mix	Val Kappa	Test Kappa
1	Inception-V3 [13]	ImageNet	×	0.8366	0.7445
2	ResNeSt-50 [18]		×	0.8214	Unknown
3	ViT-t [2]		×	0.8433	0.7860
4	ViT-s [2]		×	0.8486	0.8365
5	ViT-s [2]	EyePACS & DDR	0.0	0.8599	0.8539
6			0.1	0.8656	0.8639
7			0.5	0.8624	**0.8693**
8			1.0	0.8625	0.8427
13	TTA (t=2)	EyePACS & DDR	0.5	-	0.8636
14	TTA (t=3)			-	0.8664
15	TTA (t=4)			-	0.8592

Fundus Image Datasets. We employ two large-scale datasets, i.e. EyePACS dataset [4] and DDR dataset [6], which offer color fundus images for DR grading. Each image is categorized into five grades, including No DR, Mild NPDR, Moderate NPDR, Severe NPDR, and PDR. We adopt training sets of the two datasets, composed of 35,126 and 6,260 images for network pre-training.

4.3 Implementation Details

All images are resized to 384×384 as the input for ViT, and 512×512 for other models. Random cropping, flipping, rotation, and color jittering are applied on the training data. All images are normalized by mean=0.5 and std=0.5. We train the networks for 100 epochs using the SGD algorithm. The initial learning rate is set to 1e-3, and decay by cosine annealing learning rate schedule.

4.4 Ablation Study

In this part, we conduct an ablation study to investigate the impact of each component for DR grading, including backbone network, pre-training scheme, hybrid Mixup and CutMix strategy, and test time augmentation.

Backbone Network. As can be seen in Table 7, we compare the performance of different networks, including Inception-V3 [13], ResNeSt-50 [18], ViT-t [2], and ViT-s [2]. Among the four networks, ViT-s shows the best performance with 0.8486 val Kappa and 0.8365 test Kappa. The results demonstrate the ability of ViT to capture long-range spatial correlations within each image. Therefore, we choose ViT-s as our backbone network for DR grading.

Table 8. The top-5 results on the leaderboard of diabetic retinopathy grading.

Rank	Team	AUC	Kappa
1	FAI	0.9147	0.8910
2	KT_Bio_Health	0.9257	0.8902
3	LaTIM	0.8938	0.8761
4	BUAA_Aladinet	0.9188	0.8721
5	FDVTS_DR (ours)	0.9239	0.8636

Pre-training Scheme. As the number of OCTA images on DRAC dataset is relatively limited, we additionally adopt EyePACS [4] and DDR [6] datasets to pre-train the networks, which provide a large number of color fundus images for DR drading. Compared to the ImageNet pre-trained ViT-s (4th row), the EyePACS & DDR pre-trained ViT-s (5th row) significantly improves the performance by 1.13% val Kappa and 1.74% test Kappa. The results illustrate that the representation of color fundus images can serve as a useful substrate for DR grading on OCTA images.

Hybrid MixUp and CutMix Strategy. Moreover, the hybrid MixUp and CutMix strategy also greatly boosts the DR grading performance, as shown in the 6th to 8th rows of Table 7. We set different values of mix probability, which control the probability of a sample being mixed. When the probability is set to 0.1 (6th row), the ViT-s achieves 0.57% and 1% improvements on val and test Kappa, respectively. With the increase of probability, the test Kappa continues to improve and reaches 0.8693 when the probability is 0.5. However, if the value of probability is set too large (e.g. 1.0), the training phase would be difficult to converge, which leads to inferior grading performance (i.e. 0.8427 test Kappa). According to the results, we choose the best-performing ViT-s with the mix probability 0.5 as our final model.

Test Time Augmentation. Aiming to further boost the generalization ability of our model on the testing set, we employ a test time augmentation (TTA) operation. For each image x, we apply t-time augmentations $\mathcal{A}$ and obtain $\{\mathcal{A}_1(x), \mathcal{A}_2(x), ..., \mathcal{A}_t(x)\}$. Then, we average the model's predictions on these samples. The last three rows of Table 7 show the results of TTA with different t (i.e. $t = 2, 3, 4$). However, it is observed that our augmentations do not further promote the grading performance. It is probably because our strong augmentation is beneficial to the training phase but may be harmful in testing phase.

4.5 Results on the Leaderboard

Table 8 presents the top-5 results of our method and other participants on the testing set of task 3 in the DRAC challenge. Our EyePACS & DDR pre-trained

ViT-s model with Mix and TTA strategies ranked 5th on the leaderboard, approaching 0.9239 AUC and 0.8636 Kappa.

5   Conclusion

In this work, we presented our solutions of three tasks in the Diabetic Retinopathy Analysis Challenge. In task 1, we adopted UNet and UNet++ networks with pre-trained encoders for DR lesion segmentation, which benefited from a series of strong data augmentation. In task 2, we created an ensemble of various networks for image quality assessment. Both the pre-training scheme and hybrid MixUp and CutMix strategy helped to boost the performance. In task 3, we developed a ViT model for DR grading. We found the model pre-trained on color fundus images serves as a useful substrate for OCTA images. The experimental results demonstrated the effectiveness of our proposed methods. On the three challenge leaderboards, our methods ranked 4th, 3rd, and 5th, respectively.

References

1. Dai, L., et al.: A deep learning system for detecting diabetic retinopathy across the disease spectrum. Nat. Commun. **12**(1), 1–11 (2021)
2. Dosovitskiy, A., et al.: An image is worth 16x16 words: Transformers for image recognition at scale. arXiv preprint arXiv:2010.11929 (2020)
3. Hu, J., Shen, L., Sun, G.: Squeeze-and-excitation networks. In: Proceedings of the IEEE Conference on Computer Vision And Pattern Recognition, pp. 7132–7141 (2018)
4. Kaggle: Kaggle diabetic retinopathy detection competition. https://www.kaggle.com/c/diabetic-retinopathy-detection
5. Li, T., et al.: Applications of deep learning in fundus images: a review. Med. Image Anal. **69**, 101971 (2021)
6. Li, T., Gao, Y., Wang, K., Guo, S., Liu, H., Kang, H.: Diagnostic assessment of deep learning algorithms for diabetic retinopathy screening. Inf. Sci. **501**, 511–522 (2019)
7. Liu, R., et al.: Deepdrid: Diabetic retinopathy—grading and image quality estimation challenge. Patterns, p. 100512 (2022)
8. Ronneberger, O., Fischer, P., Brox, T.: U-Net: convolutional networks for biomedical image segmentation. In: Navab, N., Hornegger, J., Wells, W.M., Frangi, A.F. (eds.) MICCAI 2015. LNCS, vol. 9351, pp. 234–241. Springer, Cham (2015). https://doi.org/10.1007/978-3-319-24574-4_28
9. Sheng, B.,et al.: An overview of artificial intelligence in diabetic retinopathy and other ocular diseases. Front. Public Health. **10** (2022). https://doi.org/10.3389/fpubh.2022.971943
10. Sheng, B., et al.: Diabetic retinopathy analysis challenge 2022, March 2022. https://doi.org/10.5281/zenodo.6362349
11. Spaide, R.F., Fujimoto, J.G., Waheed, N.K., Sadda, S.R., Staurenghi, G.: Optical coherence tomography angiography. Prog. Retin. Eye Res. **64**, 1–55 (2018)
12. Szegedy, C., Ioffe, S., Vanhoucke, V., Alemi, A.A.: Inception-v4, inception-resnet and the impact of residual connections on learning. In: Thirty-first AAAI Conference on Artificial Intelligence (2017)

13. Szegedy, C., Vanhoucke, V., Ioffe, S., Shlens, J., Wojna, Z.: Rethinking the inception architecture for computer vision. In: Proceedings of the IEEE Conference on Computer Vision and Pattern Recognition, pp. 2818–2826 (2016)

14. Tan, M., Le, Q.: Efficientnet: rethinking model scaling for convolutional neural networks. In: International Conference on Machine Learning, pp. 6105–6114. PMLR (2019)

15. Wang, Y., et al.: A deep learning-based quality assessment and segmentation system with a large-scale benchmark dataset for optical coherence tomographic angiography image (2021)

16. Xie, S., Girshick, R., Dollár, P., Tu, Z., He, K.: Aggregated residual transformations for deep neural networks. In: Proceedings of the IEEE Conference on Computer Vision and Pattern Recognition, pp. 1492–1500 (2017)

17. Yun, S., Han, D., Oh, S.J., Chun, S., Choe, J., Yoo, Y.: Cutmix: regularization strategy to train strong classifiers with localizable features. In: Proceedings of the IEEE/CVF International Conference on Computer Vision, pp. 6023–6032 (2019)

18. Zhang, H., et al.: Resnest: split-attention networks. In: Proceedings of the IEEE/CVF Conference on Computer Vision and Pattern Recognition, pp. 2736–2746 (2022)

19. Zhang, H., Cisse, M., Dauphin, Y.N., Lopez-Paz, D.: mixup: Beyond empirical risk minimization. arXiv preprint arXiv:1710.09412 (2017)

20. Zheng, Y., He, M., Congdon, N.: The worldwide epidemic of diabetic retinopathy. Indian J. Ophthalmol. **60**(5), 428 (2012)

21. Zhou, Z., Rahman Siddiquee, M.M., Tajbakhsh, N., Liang, J.: UNet++: a nested u-net architecture for medical image segmentation. In: Stoyanov, D., et al. (eds.) DLMIA/ML-CDS -2018. LNCS, vol. 11045, pp. 3–11. Springer, Cham (2018). https://doi.org/10.1007/978-3-030-00889-5_1

Deep Learning-Based Multi-tasking System for Diabetic Retinopathy in UW-OCTA Images

Jungrae Cho[1], Byungeun Shon[1,2], and Sungmoon Jeong[1,2(✉)]

[1] Research Center for Artificial Intelligence in Medicine,
Kyungpook National University Hospital, Daegu, South Korea
jeongsm00@gmail.com
[2] Department of Medical Informatics, School of Medicine,
Kyungpook National University, Daegu, South Korea

Abstract. Diabetic retinopathy causes various abnormality in retinal vessels. In addition, Detection and identification of vessel anomaly are challenging due to nature of complexity in retinal vessels. UW-OCTA provides high-resolution image of those vessels to diagnose lesions of vessels. However, the image suffers noise of image. We here propose a deep learning-based multi-tasking systems for DR in UW-OCTA images to deal with diagnosis and checking image quality. We segment three kinds of retinal lesions with data-adaptive U-Net architectures, i.e. nnUNet, grading images on image quality and DR severity grading by soft-voting outputs of fine-tuned multiple convolutional neural networks. For three tasks, we achieve Dice similarity coefficient of 0.5292, quadratic weighted Kappa of 0.7246, and 0.7157 for segmentation, image quality assessment, and grading DR for test set of DRAC2022 challenge. The performance of our proposed approach demonstrates that task-adaptive U-Net planning and soft ensemble of CNNs can provide enhancement of the performance of single baseline models for diagnosis and screening of UW-OCTA images.

Keywords: SS-OCTA · UW-OCTA · diabetic retinopathy · semantic segmentation · ensemble

1 Introduction

Blindness has several causes from retinal diseases, one of the major factors is diabetic retinopathy (DR) with long-term experience of diabetes over 15 years [1]. As DR progresses, it generates abnormality in retinal vessels, e.g. intraretinal microvascular abnormalities (IRMAs), nonperfusion areas, and neovascularization. Since identification of those biomarkers is important to diagonose DR, several image modalities have been used in diagnosis process, e.g. color fundus images [2–4], optical coherence tomography angiography (OCTA) [5], swept-source OCTA (SS-OCTA) [6,7], and ultra-wide OCTA (UW-OCTA) [8]. Among them, UW-OCTA has been recently utilized to diagnose DR with its higher resolution than

B. Sheng and M. Aubreville (Eds.): MIDOG 2022/DRAC 2022, LNCS 13597, pp. 88–96, 2023.
https://doi.org/10.1007/978-3-031-33658-4_9

basic OCTA [9]. However, there has been no work on automatic DR grading and detection of vessel anomaly using those ultra-wide retinal vessel images [10]. To address and solve these problems, Diabetic Retinopathy Analysis Challenge in 2022 [10] has released UW-OCTA dataset for DR with three tasks, which are semantic segmentation of vessel abnormalities, grading quality of UW-OCTA images, and grading DR severity in three classes respectively. To solve each task, we propose a deep learning-based multi-tasking system for DR in retinal vessel images. We train a semantic segmentation model to segment lesions (IRMAs, non-perfusion areas, and neovascularization) with data-adaptive U-net [11] architectures, i.e. nnUNet [12]. For second and final tasks, we grade OCTA images on by ensemble of soft outputs from several validated convolutional neural networks (CNNs) by fine-tuning them respectively. The proposed method achieves Dice similarity coefficient (DSC) of 0.5292 in the first task, quadratic weighted Kappa of 0.7246, and 0.7157 in the second and final tasks. These experimental results show that adaptive planning of U-Net architecture and soft-voting of CNNs can enhance the performance of baseline models, which are basic U-Net architecture and single CNN models.

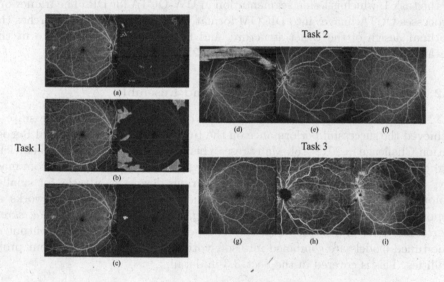

Fig. 1. Examples of datasets for task 1, 2, and 3. In the task 1, the left column indicates original images, and the other is overlay of segmentation mask (bright area) with (a) IRMAs, (b) nonperfusion areas, and (c) neovascularization. The task 2 shows (d) poor, (e) good, and (f) excellent quality level. In the task 3, there are images for (g) normal, (h) NPDR, and (i) PDR.

2   Related Works

2.1   Semantic Segmentation

Semantic segmentation based on deep learning has contributed in medical imaging field. U-net [11] is one of the representative architecture of deep learning-based segmentation, which preserves local information efficiently with its skip connections between features of encoder and decoder paths. Variants of U-net architecture have been applied to various modalities, e.g. MRI, CT, OCT, and even natural images. F. Milletari et al. (2016) [13] proposed V-Net segmenting three dimensional volumetric data. Zhou et al. (2018) [14] suggested nested and dense skip connections with deep supervision in the U-net architecture. Unet 3+ [15] extended skip connections into full-scale features in the U-net. However, even enhanced variants drive the performance of segmentation higher, the diversity of medical image dataset still requires for researchers to fit the hyper-paameters of those architectures empirically. nnUNet [12] provided automated planning for the U-net architecture on given domain and dataset. To optimize the architecture of U-net, we here select the nnUNet framework as a start point of the task 1, which is lesion segmentation in UW-OCTA for DR. The framework processes OCTA images into DICOM format, crop them randomly, searches the optimal design of the U-net structure, and trains the model finally to fit the dataset. Details are described in the Sect. 3.2.

2.2   Convolutional Neural Network and Ensemble

CNN has proved its performance on image data over the past decade since it achieved the successful performance on the ImageNet Large Scale Visual Recognition Challenge in 2012 [16]. Many researchers have devised deeper and efficient structures of CNN, and released pretrained weights of those networks from massive dataset to allow fine-tuning them to various downstream tasks for smaller amount of dataset. In this paper, we gather several pretrained networks as reported in timm [17] including *resnet50d, efficientnetb3, efficientnetb3a, skresnext50_32 × 4d, and seresenext50d_32 × 4d* empirically [18–23]. The output of fine-tuned models are combined into soft-voting [24] by averaging output probabilities. This is covered in the Sect. 3.3 in detail.

3   Methods

3.1   Dataset

Dataset released by DRAC 2022 Challenge includes sub-datasets for each task. All the data are UW-OCTA images acquired from SS-OCTA system VG200D by SVision Imaging, Ltd. in China. The size of images are 1024×1024 with PNG format. The dataset for the task 1 contains 109 images and their labels. There are three kinds of lesion labels in the dataset, which are IRMAs, nonperfusion areas, and neovascularization. The second dataset is for the task 2, i.e. image quality

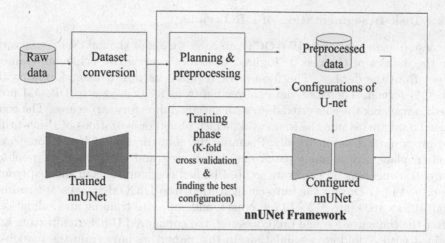

Fig. 2. Workflow of the nnUNet framework.

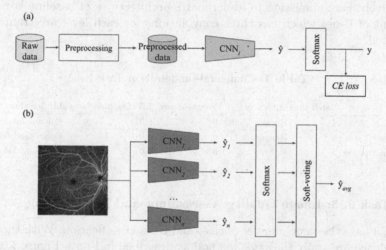

Fig. 3. Overview of training and inference of soft-ensemble of multiple CNNs. (a) is a training phase for each CNN_i, and (b) is inference phase of mutlipe CNNs with soft-voting.

levels with poor quality, good quality, and excellent quality levels. The dataset has 665 images and matching labels. The final dataset for the task 3 consists of 611 images and labels with three DR gradings, normal, non-proliferatived diabetic retinopathy (NPDR), and proliferatived diabetic retinopathy (PDR). For the competition, organizers of the challenge also releases test set, which are 65, 438, and 386 images for task 1, 2, and 3 without ground-truth labels. Figure 1 shows example of images and labels for each task.

3.2 Task 1: Segmentation of DR Lesions

For lesion segmentation in UW-OCTA images, we adopt the nnUNet framework for the dataset of the task 1. Figure 2 describes overview of achieving trained U-net from the dataset. The framework requires its own dataset format with DICOM format. Therefore, data conversion from RGB images to DICOM preceeds, arranging the converted data with specified directory structures. The converted data are fed into the framework, which plans configurations of U-net fit in the given dataset automatically. Planning the configuration of U-net comprises median plane shape, input patch size, batch size, and the number of pooling per axis, depending on hardware setup. Planned configurations in this paper are described in 1. During the auto-configuration, the DICOM images are resized from 1024×1024 into 512×512, cropped, and split into training and validation. After the configuration and preprocessing, the configured U-net architecture are trained with K-fold cross-validation. In this paper, we only train the model in one fold ($K=1$) with 20 epochs. We train nnUNets for each segmentation label respectively. For submission to leaderboard, architectures of baseline U-net and a variant of U-net which has three convolutions for each level are trained and submitted.

Table 1. Configuration details of Task 1.

Sub-task	Batch size	Optimizer	Weight deacy	LR (Learning rate)	LR Scheduler
IRMAs	4	Ranger [25, 26]	0	0.01	Polynomial decay [27]
Nonperfusion area	5	Ranger	0	0.01	Polynomial decay
Neovascularization	2	Ranger	0	0.01	Polynomial decay

3.3 Task 2, 3: Image Quality Assessment and DR Grading

Task 2, 3 have the same number of class and task, classification. With those correspondence, we apply the same method as described in Fig. 3. Figure 3 consists of two stages. The stage (a) depicts training CNN models. As mentioned in the section of Introduction, we select multiple different CNN structures, which has been pretrained on ImageNet. The list of networks are *resnet50d, efficientnetb3, efficientnetb3a, skresnext50_32×4d, and seresenext50d_32×4d*. These models are chosen based on their performances presented in the document of timm [17]. As shown in Fig. 3 (a), raw data are preprocessed with resizing from 1024×1024 into 224×224 (*resnet50d and seresenext50d_32×4d*), 288×288 (*skresnext50_32×4d*), 300×300 (*efficientnetb3*), and 320×320 (*efficientnetb3a*), normalization, and randomly vertical or horizontal flip of images. Table 2 shows implementation details on each model for both Task 2 and 3. In table 3, image size indicates both height and width of image size, and LR is abbreviation of learning rate. There was no LR scheduler for Task 2 and 3. Each $CNN_i (i = [1, n], n = 5)$ is trained on the preprocessed images with cross entropy loss between the ground-truth label y and the output vector of $\hat{y}$ after softmax operation. All the trained models in Fig. 3 (a) are combined with ensemble of outputs by averaging softmax outputs of models, i.e. soft-voting as shown in Fig. 3 (b). Training pipeline is implemented with Lightning [28].

Table 2. Implementation details of models for Task 2 and 3.

Model	Image size	Batch size	Optimizer	Weight decay	LR
resnet50d	224	32	Adam [29]	0	0.001
efficientnetb3	320	32	Adam	0	0.001
efficientnetb3a	300	32	Adam	0	0.001
skresnext50_32 × 4d	288	32	Adam	0	0.001
seresenext50d_32 × 4d	224	32	Adam	0	0.001

4 Experiments

In this section, we evaluate the performance of lesion segmentation, image quality assessment, and DR grading on the basis of leader-board in DRAC 2022 Challenge. All the tables in the paper contains the performance of top-3 ranks and our methods. Table 3 shows averaged Dice Similarity Coefficient (DSC) and intersection over union (IoU) of models. For nnUNets, the variant recorded higher performances than the baseline nnUNet, achieving 21-th rank in the leaderboard. Table 4 shows the quadratic weighted Kappa and area under curve (AUC) on task 2 and 3. For the task 2, we recorded 41-th rank with the soft-voting method from 5 CNNs. The same methods ranked in 74-th entry. Both tasks demonstrate

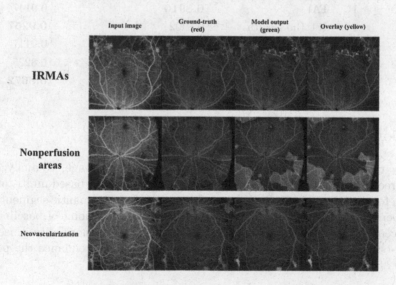

Fig. 4. Visualization of segmentation results from validation set.

that soft-voting from 5 CNN models enhanced performances than single CNN models as described in Table 4. For the lesion segmentation task, we visualize segmentation results by the nnUNet variant qualitatively as shown in Fig. 4. In Fig. 4, there are input images, ground-truth labels (red), outputs of the model (green), and overlay of both segmentation masks (yellow). Those visualizations are from validation set split from training set released by DRAC 2022 challenge.

Table 3. The performance of the task 1.

Rank	Team (model)	mean DSC	mean IoU
1	FAI	**0.6067**	**0.4590**
2	KT Bio Health	0.6046	0.4573
3	KT Bio Health	0.6044	0.4579
21	Ours (nnUNet)	0.4942	0.3605
	Ours (nnUNet variant)	**0.5292**	**0.3889**

Table 4. The performance of the task 2 and 3.

Task	Rank	Team (model)	Quadratic Weighted Kappa	AUC
2	1	FAI	**0.8090**	0.8238
	2	KT Bio Health	0.8075	**0.9085**
	3	FAI	0.8044	0.8173
2	41	Ours (efficientnet_b3)	0.6559	**0.8765**
		Ours (soft-voting)	**0.7246**	0.84
3	1	FAI	**0.8910**	0.9147
	2	KT Bio Health	0.8902	**0.9257**
	3	FAI	0.8876	0.9117
3	74	Ours (resnet50d)	0.657	0.8225
		Ours (soft-voting)	**0.7157**	**0.8672**

5 Conclusion

Automatic DR analysis in UW-OCTA is in the pioneering stage. To analyze DR and screen the quality of images, we propose a deep learning-based multi-tasking system for the DR analysis. Planning architecture of deep semantic segmentation model enables segmentation of DR lesions. It can be start point of baseline and enhancement of semantic segmentation for retinal vessel abnormality. Ensemble of soft-outputs of multiple CNNs demonstrates that it can enhance the performance of generalization for unseen test data.

Code availability

Implementations in this paper will be released here soon: https://github.com/jcho2022/drac2022_jcho.

Acknowledgements. This research was partly supported by a grant of the Korea Health Technology R&D Project through the Korea Health Industry Development Institute (KHIDI), funded by the Ministry of Health & Welfare, Republic of Korea (grant number : HR22C1832) and the MSIT(Ministry of Science and ICT), Korea, under the ITRC(Information Technology Research Center) support program(IITP-2022-2020-0-01808) supervised by the IITP(Institute of Information & Communications Technology Planning & Evaluation).

References

1. Tian, M., Wolf, S., Munk, M.R., Schaal, K.B.: Evaluation of different swept'source optical coherence tomography angiography (ss-octa) slabs for the detection of features of diabetic retinopathy. Acta ophthalmologica **98**(4), e416–e420 (2020)
2. Dai, L., et al.: A deep learning system for detecting diabetic retinopathy across the disease spectrum. Nat. Commun. **12**(1), 1–11 (2021)
3. Liu, R., et al.: Deepdrid: diabetic retinopathy-grading and image quality estimation challenge. Patterns **3**(6), 100512 (2022)
4. Sheng, B., et al.: An overview of artificial intelligence in diabetic retinopathy and other ocular diseases. Front. Public Health **10**, 971943 (2022)
5. Spaide, R.F., Fujimoto, J.G., Waheed, N.K., Sadda, S.R., Staurenghi, G.: Optical coherence tomography angiography. Progress Retinal Eye Res. **64**, 1–55 (2018)
6. Schaal, K.B., Munk, M.R., Wyssmueller, I., Berger, L.E., Zinkernagel, M.S., Wolf, S.: Vascular abnormalities in diabetic retinopathy assessed with swept-source optical coherence tomography angiography widefield imaging. Retina **39**(1), 79–87 (2019)
7. Stanga, P.E., et al.: New findings in diabetic maculopathy and proliferative disease by swept-source optical coherence tomography angiography. OCT Angiography Retinal Macular Dis. **56**, 113–121 (2016)
8. Zhang, Q., Rezaei, K.A., Saraf, S.S., Chu, Z., Wang, F., Wang, R.K.: Ultra-wide optical coherence tomography angiography in diabetic retinopathy. Quant. Imaging Med. Surgery **8**(8), 743 (2018)
9. Russell, J.F., Shi, Y., Hinkle, J.W., Scott, N.L., Fan, K.C., Lyu, C., Gregori, G., Rosenfeld, P.J.: Longitudinal wide-field swept-source oct angiography of neovascularization in proliferative diabetic retinopathy after panretinal photocoagulation. Ophthalmol. Retina **3**(4), 350–361 (2019)
10. Sheng, B., et al.: Diabetic retinopathy analysis challenge 2022, March 2022. https://doi.org/10.5281/zenodo.6362349
11. Ronneberger, O., Fischer, P., Brox, T.: U-net: convolutional networks for biomedical image segmentation. In: Navab, N., Hornegger, J., Wells, W.M., Frangi, A.F. (eds.) MICCAI 2015. LNCS, vol. 9351, pp. 234–241. Springer, Cham (2015). https://doi.org/10.1007/978-3-319-24574-4_28
12. Isensee, F., Jaeger, P.F., Kohl, S.A., Petersen, J., Maier-Hein, K.H.: nnu-net: a self-configuring method for deep learning-based biomedical image segmentation. Nat. Methods **18**(2), 203–211 (2021)

13. Milletari, F., Navab, N., Ahmadi, S.A.: V-net: fully convolutional neural networks for volumetric medical image segmentation. In: 2016 Fourth International Conference on 3D Vision (3DV), pp. 565–571. IEEE (2016)

14. Zhou, Z., Rahman Siddiquee, M.M., Tajbakhsh, N., Liang, J.: UNet++: a nested u-net architecture for medical image segmentation. In: Stoyanov, D., et al. (eds.) DLMIA/ML-CDS -2018. LNCS, vol. 11045, pp. 3–11. Springer, Cham (2018). https://doi.org/10.1007/978-3-030-00889-5_1

15. Huang, H., et al.: Unet 3+: a full-scale connected unet for medical image segmentation. In: ICASSP 2020–2020 IEEE International Conference on Acoustics, Speech and Signal Processing (ICASSP), pp. 1055–1059. IEEE (2020)

16. Russakovsky, O., et al.: ImageNet large scale visual recognition challenge. Int. J. Comput. Vis. (IJCV) **115**(3), 211–252 (2015). https://doi.org/10.1007/s11263-015-0816-y

17. Wightman, R.: Pytorch image models. https://github.com/rwightman/pytorch-image-models (2019). https://doi.org/10.5281/zenodo.4414861

18. He, T., Zhang, Z., Zhang, H., Zhang, Z., Xie, J., Li, M.: Bag of tricks for image classification with convolutional neural networks. In: Proceedings of the IEEE/CVF Conference on Computer Vision and Pattern Recognition, pp. 558–567 (2019)

19. Tan, M., Le, Q.: Efficientnet: Rethinking model scaling for convolutional neural networks. In: International conference on machine learning. pp. 6105–6114. PMLR (2019)

20. Xie, S., Girshick, R., Dollár, P., Tu, Z., He, K.: Aggregated residual transformations for deep neural networks. In: Proceedings of the IEEE Conference on Computer Vision and Pattern Recognition, pp. 1492–1500 (2017)

21. Li, X., Wang, W., Hu, X., Yang, J.: Selective kernel networks. In: Proceedings of the IEEE/CVF Conference on Computer Vision and Pattern Recognition, pp. 510–519 (2019)

22. Lee, J., Won, T., Lee, T.K., Lee, H., Gu, G., Hong, K.: Compounding the performance improvements of assembled techniques in a convolutional neural network. arXiv preprint arXiv:2001.06268 (2020)

23. Hu, J., Shen, L., Sun, G.: Squeeze-and-excitation networks. In: Proceedings of the IEEE Conference on Computer Vision and Pattern Recognition, pp. 7132–7141 (2018)

24. Peppes, N., Daskalakis, E., Alexakis, T., Adamopoulou, E., Demestichas, K.: Performance of machine learning-based multi-model voting ensemble methods for network threat detection in agriculture 4.0. Sensors **21**(22), 7475 (2021)

25. Liu, L., et al.: On the variance of the adaptive learning rate and beyond. arXiv preprint arXiv:1908.03265 (2019)

26. Zhang, M., Lucas, J., Ba, J., Hinton, G.E.: Lookahead optimizer: k steps forward, 1 step back. In: Advances in Neural Information Processing Systems, vol. 32 (2019)

27. Chen, L.C., Papandreou, G., Schroff, F., Adam, H.: Rethinking atrous convolution for semantic image segmentation. arXiv preprint arXiv:1706.05587 (2017)

28. Falcon, W., The PyTorch lightning team: PyTorch lightning, March 2019. https://doi.org/10.5281/zenodo.3828935, https://github.com/Lightning-AI/lightning

29. Kingma, D.P., Ba, J.: Adam: a method for stochastic optimization. arXiv preprint arXiv:1412.6980 (2014)

Semi-supervised Semantic Segmentation Methods for UW-OCTA Diabetic Retinopathy Grade Assessment

Zhuoyi Tan[ID], Hizmawati Madzin[✉][ID], and Zeyu Ding[ID]

Faculty of Computer Science and Information Technology, Universiti Putra Malaysia,
Serdang 43400, Malaysia
hizmawati@upm.edu.my

Abstract. People with diabetes are more likely to develop diabetic retinopathy (DR) than healthy people. However, DR is the leading cause of blindness. At present, the diagnosis of diabetic retinopathy mainly relies on the experienced clinician to recognize the fine features in color fundus images. This is a time-consuming task. Therefore, in this paper, to promote the development of UW-OCTA DR automatic detection, we propose a novel semi-supervised semantic segmentation method for UW-OCTA DR image grade assessment. This method, first, uses the MAE algorithm to perform semi-supervised pre-training on the UW-OCTA DR grade assessment dataset to mine the supervised information in the UW-OCTA images, thereby alleviating the need for labeled data. Secondly, to more fully mine the lesion features of each region in the UW-OCTA image, this paper constructs a cross-algorithm ensemble DR tissue segmentation algorithm by deploying three algorithms with different visual feature processing strategies. The algorithm contains three sub-algorithms, namely pre-trained MAE, ConvNeXt, and SegFormer. Based on the initials of these three sub-algorithms, the algorithm can be named MCS-DRNet. Finally, we use the MCS-DRNet algorithm as an inspector to check and revise the results of the preliminary evaluation of the DR grade evaluation algorithm. The experimental results show that the mean dice similarity coefficient of MCS-DRNet v1 and v2 are 0.5161 and 0.5544, respectively. The quadratic weighted kappa of the DR grading evaluation is 0.7559. Our code is available at https://github.com/SupCodeTech/DRAC2022.

Keywords: Semantic segmentation · UW-OCTA image ·
Semi-supervised learning · Deep learning

1 Introduction

Diabetic retinopathy (DR) grade assessment is a very challenging task. Because in this work, it is always difficult for us to capture some very subtle DR feature

Supported by Ministry of Higher Education, Malaysia.

in the image. However, capturing these subtle features is extremely important for assessing the severity of diabetic retinopathy. The traditional assessment of DR grading mainly relies on fundus photography and FFA, especially for PDR (proliferatived diabetic retinopathy) and NPDR (non-proliferatived diabetic retinopathy). The main drawback òf these tests is that they can cause serious harm to vision health. In addition, the FA detection method is mainly used to detect the presence or absence of neovascularization. It is often difficult to detect early or small neovascularization lesions by simply relying on fundus photography. Second, the test is an invasive fundus imaging method, which cannot be used in patients with allergies, pregnancy, and poor liver and kidney function. Therefore, for most cases, DR detection mainly adopts the ultra-wide OCTA method. This method can noninvasively detect the changes of DR neovascularization, which is an important imaging method for the ophthalmic diagnosis of PDR. However, at present, due to there being no work on automated DR analysis using standard ultra-wide (swept-source) optical coherence tomography angiography (UW-OCTA) [7,12,17,18], the available UW-OCTA datasets are scarce and training data with label information is limited.

In recent years, to solve the problem of lack of labeled data, researchers have proposed many methods to alleviate the need for labeled data [2,4,8,19,23]. Among these methods, self-supervised learning is one of the most representative methods [1,3,4,11,22]. Self-supervised learning is mainly through pre-training on a large number of unlabeled data, and then mining the supervised features hidden in the unlabeled data. For example, the masked image modeling algorithm [1,10] in the domain of self-supervised learning. The algorithm randomly masks a part of the image in different ways. And then let the algorithm learn on its own how to recover the broken area. This method can make the algorithm learn useful algorithm weights for downstream tasks (classification [15,21], semantic segmentation [14,16], etc.) during the pre-training period, so as to improve the recognition accuracy of the algorithm in downstream tasks.

Nowadays, self-supervised learning has been widely used in various fields and has achieved good results [26]. However, the fly in the ointment is that in the process of self-supervised learning, it is difficult for an algorithm to efficiently dig out the supervised information beneficial to downstream tasks from a completely unfamiliar field without relying on any hints and only relying on its own intuition. For example, when we need to organize some unfamiliar and disorganized documents, we always find it very difficult. Because there are too many uncertainties here. However, if someone gives us information related to the classification of these documents. This may be of great help to us because it gives us a first glimpse of these unfamiliar files. To this end, in this paper, in order to mine supervised features contained in UW-OCTA images more efficiently, we construct a method between supervised learning and self-supervised learning, Briefly, this method is a suite of semi-supervised semantic segmentation methods for diabetic retinopathy grade assessment. Overall, the method can be divided into the following two frameworks:

The first framework is a semantic segmentation method of UW-OCTA diabetic retinopathy image with semi-supervised cross-algorithm ensemble. In this framework, first, we pre-train the MAE algorithm [10] on the DR grade assessment dataset in the DRAC2022 challenge. MAE algorithm [10] as a common structure in self-supervised masked image modeling visual representation learning [1,10]. The algorithm can alleviate the labeling requirements of UW-OCTA images through self-supervised pre-training. Although the MAE [10] algorithm designed based on the vision transformer (ViT) [9] architecture has good performance, the algorithm of this architecture outputs single-scale low-resolution features instead of multi-scale features. However, by fusing multi-scale UW-OCTA image features, our method can better cope with the problem of reduced recognition accuracy caused by the size change of the lesion area. To this end, we also deploy a no-position encoding and layered transformer encoder, SegFormer [24]. The SegFormer [24] algorithm is good at mining the features of the lesion area in the image at different scales. On the other hand, the lightweight All-MLP decoder design adopted by SegFormer [24] can generate powerful representations without the need for complex and computationally demanding modules. Finally, in order to form a good complement with MAE [10] and SegFormer algorithm [24] based on the self-attention mechanism. We also integrate an algorithm for pure convolutional architecture based on the "sliding window" strategy (i.e., without self-attention), ConvNeXt [13]. The introduction of the convolution architecture makes our method more comprehensively obtain the global features (long-distance dependencies) in UW-OCTA images. Therefore, in this framework, we construct a semi-supervised cross-algorithm ensemble method for lesion identification in UW-OCTA DR images based on MAE, ConvNeXt, and SegFormer algorithms [5,6,10,13,24]. Based on the initials of these three sub-algorithms, our method can be named *MCS-DRNet*. There are two versions of our method in total, MCS-DRNet v1 and v2, corresponding to the challenge version and the post-challenge version, respectively.

The second framework is a grading assessment method for diabetic retinopathy based on a threshold inspection mechanism. In this framework, we use the MCS-DRNet as an inspector to check and revise the results of the preliminary evaluation of the DR grade evaluation algorithm (EfficientNet v2 [20]). To be specific, the inspector is mainly composed of the DR region segmentation pre-trained weights of MCS-DRNet and grading evaluation algorithm. In addition, we set up a set of threshold systems and evaluation rules for evaluating the severity of DR.

Finally, on the test set of the diabetic retinopathy analysis challenge (DRAC) 2022, our method exhibits strong performance. In terms of DR semantic segmentation, the mean dice similarity coefficient of MCS-DRNet v1 and MCS-DRNet v2 are 0.5161 and 0.5544, respectively. In terms of DR grading evaluation, the quadratic weighted kappa of the DR grading evaluation method based on the inspection mechanism is 0.7559, which brings a 2.99% improvement compared to the baseline method (EfficientNet v2 [20]).

2   Approach

2.1   MCS-DRNet Series Methods

The series method of MCS-DRNet we built can be divided into two stages, as shown in Fig. 1. These two phases are responsible for accomplishing two different tasks. The first stage is the pre-task training. This stage mainly relies on the MAE algorithm [10] to pre-train on the diabetic retinopathy grade classification data set (Task 3) in the DRAC2022 challenge. In this process, for the MAE pre-training algorithm to comprehensively mine and learn the supervision information contained in the UW-OCTA image, the algorithm performs a masking operation on most of the areas in the input UW-OCTA image. The specific masking effect is shown in Fig. 2.

Then, the MAE encoder [10] performs feature extraction on the visible patches subset of the image. In the process, the encoder will become a series of mask tokens. The full encoded patches and mask tokens are then processed by a small decoder. The main function of this small decoder is to reconstruct the original image in pixels. After the pre-training, we would extract the backbone of the MAE encoder (also considered as pre-trained MAE-ViT) [9,10]. And then, we use this extracted MAE-ViT as the initialization weight for the semantic segmentation task of the MAE algorithm. However, the MAE decoder is discarded [10].

The second stage is semantic segmentation learning (supervised learning) of UW-OCTA images. In this stage, in order to more fully explore the lesion features regions in UW-OCTA images, three algorithms with different visual processing strategies are deployed in the proposed method, which are pre-traind MAE [10], ConvNeXt [13], and SegFormer [24]. In this stage of learning, in order to speed up the whole training process, we set the training image size of the MAE [10] and ConvNeXt [13] algorithm to 640×640.

SegFormer [24], a hierarchical transformer without position coding. Because the algorithm does not adopt the interpolation structure of position coding in the ViT algorithm, the algorithm can well cope with the influence brought by the recognition process of lesions with different sizes. Moreover, the high-resolution fine features and low-resolution coarse features generated by the encoder in Seg-Former architecture can make up for the shortcoming that the MAE algorithm can only generate a single fixed-resolution low-resolution feature graphics to some extent. The MLP decoder part of the SegFormer algorithm [24] aggregates information from both lower and higher levels (lower levels tend to keep local features, while higher levels focus on global features). This enables the SegFormer network [24] to fuse multi-scale UW-OCTA image features in the process of processing UW-OCTA images, improve the perception field of the algorithm, and present a powerful representation.

ConvNeXt [13], a pure convolutional architecture algorithm (without Self-attention architecture). To some extent, this algorithm retains the "sliding window" strategy, which is beneficial to play the unique advantages of convolutional architecture in obtaining global image features (long-distance dependencies).

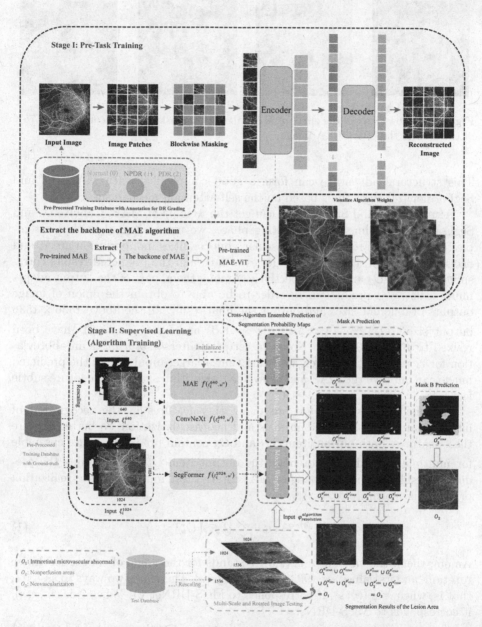

Fig. 1. The realization principle of MCS-DRNet method. Mask A indicates that intraretinal microvascular abnormalities and neovascularization are fused on a single mask for training, and Mask B indicates that the nonperfusion areas category is trained as a single mask.

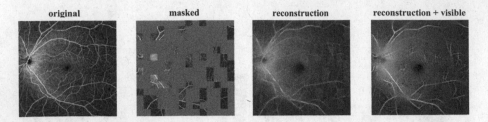

original   masked   reconstruction   reconstruction + visible

Fig. 2. The flow chart of the effect of MAE algorithm masking and reconstructing 75% of the area in the UW-OCTA image.

This structural advantage can form a good complement with MAE [10] and SegFormer [24] algorithms based on the self-attention mechanism.

After Stage I and Stage II, we can get the weights of MAE, ConvNeXt, and SegFormer algorithms. In the testing phase, we use a cross-algorithm ensemble to predict intraretinal microvascular abnormalities, nonperfusion areas, and neovascularization, the three categories of DR, and the predicted results corresponded to O_1, O_2 and O_3, respectively. Among them, part of the output adopts multi-scale (MS) prediction. In order to be able to obtain the union of image outputs of different sizes and output the final result, all the above 1536×1536 output sizes $\{O_\nu^{\varphi_{1536R}^\lambda}, (\lambda = m, c, s; \nu = 1, 3)$ and $(\lambda = c; \nu = 2)$ }have been resized to 1024×1024 size (Note: the letter R after *resolution* is an abbreviation for *rescaling*). Moreover, we found that a larger input size of the predicted image can improve the MAE and ConvNeXt algorithm's ability to capture subtle features.

In conclusion, since there are many tiny features in the recognition of intraretinal microvascular abnormals and neovascularization, to better capture these features, both output items O_1 and O_3 contain predicted values ensemble for multiple algorithms. Moreover, the output item $O_\nu^{\varphi_{1536R}^i}$ contains the prediction results of the multi-angle (MA) rotated test image, and the realization principle is shown in Eq. 1.

$$O_\nu^{\varphi_{1536R}^\lambda} = O_\nu^{\varphi_{1536R}^\lambda} \cup f_\tau \left(O_\nu^{\varphi_{1536R}^\lambda} \right) \qquad (1)$$

Among them, the value of ν is $(1,3)$. τ stands for the angle of counterclockwise rotation, and its values are $(90, 180, 270)$ degrees. There are two ways to assign i, that is, when λ is (m, s), it corresponds to MCS-DRNet v1. When λ is (m, c, s), it corresponds to MCS-DRNet v2.

In order to show the implementation principle of the formula $f_\tau \left(O_\nu^{\varphi_{1536R}^\lambda} \right)$ more concretely, we visualize how an example of this formula is implemented (in this example, $\tau = 180, \nu = 1, \lambda = c$.), as shown in Fig. 3.

For the output O_2 of the nonperfusion areas, only $O_2^{\varphi_{1536R}^c}$ is included.

According to the above method, the two versions of the MCS-DRNet method constructed in this paper are as follows:

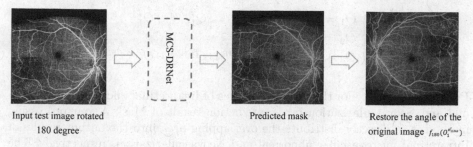

Input test image rotated 180 degree

Predicted mask

Restore the angle of the original image $f_{180}(O_1^{\varphi_{1536}^c})$

Fig. 3. The flow chart of the $f_{180}\left(O_1^{\varphi^c_{1536R}}\right)$ algorithm to rotate the image.

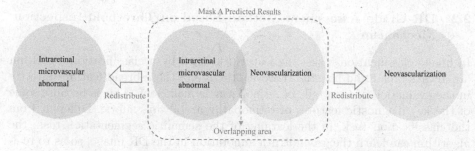

Fig. 4. The processing method of the overlapping area of Intraretinal microvascular abnormal and Neovascularization in the prediction result of Mask A.

MCS-DRNet V1 (Challenge Version). In this version of the method, the version used by the ConvNeXt algorithm is L, and the algorithm is only used to predict Mask B. SegFormer and MAE algorithms are used for the prediction of Mask A. Therefore, the output items of MCS-DRNet v1 is shown in Equation 2.

$$\begin{cases} O_1 = O_1^{\varphi^m_{1536R}} \cup O_1^{\varphi^s_{1024}} \cup O_1^{\varphi^s_{1536R}} \\ O_2 = O_2^{\varphi^{c^*}_{1536R}} \\ O_3 = O_3^{\varphi^m_{1536R}} \cup O_3^{\varphi^s_{1024}} \cup O_3^{\varphi^s_{1536R}} \end{cases} \tag{2}$$

The asterisk ($*$) in the upper right corner of the letter c indicates that the version of the algorithm ConvNeXt is L

MCS-DRNet V2 (Post-challenge Version). The method of this version is mainly based on the method of the previous version, with a slight improvement. The direction of improvement is reflected in: the version of the ConvNeXt algorithm is upgraded from L to XL. Furthermore, the algorithm is not only used for the prediction of Mask B, but also for the prediction of Mask A. Therefore, the output items of MCS-DRNet v2 is shown in Eq. 3.

$$\begin{cases} O_1 = O_1^{\varphi_{1536R}^m} \cup O_1^{\varphi_{1536R}^c} \cup O_1^{\varphi_{1024}^s} \cup O_1^{\varphi_{1536R}^s} \\ O_2 = O_2^{\varphi_{1536R}^c} \\ O_3 = O_3^{\varphi_{1536R}^m} \cup O_3^{\varphi_{1536R}^c} \cup O_3^{\varphi_{1024}^s} \cup O_3^{\varphi_{1536R}^s} \end{cases} \quad (3)$$

Post-processing. For the overlapping area of Intraretinal microvascular abnormal and Neovascularization in the prediction result of Mask A, our processing method is to directly distribute the overlapping area into the output results of Intraretinal microvascular abnormal and Neovascularization, respectively. The implementation process is shown in Fig. 4.

2.2 DR Grade Assessment Method Based on Threshold Inspection Mechanism

In image classification tasks, the features learned by a classification algorithm are always broad and holistic. However, the use of these broad features directly in the evaluation of the grade of retinopathy is inefficient. This is because some of the key diagnostic features of retinopathy are always subtle. Different from the classification task, in the learning of the semantic segmentation task, the algorithm can learn the groundtruth annotation in the DR image, so as to realize the acquisition of the detailed lesion features in the image. The limitation of semantic segmentation algorithms is that they cannot automatically classify images based on the learned features. For this reason, it seems to us that there is some kind of complementary relationship between them for the classification and segmentation tasks of UW-OCTA images. This relationship can be understood as subtle features learned in segmentation tasks can be used to examine and correct misclassified classes in classification tasks. Motivated by this idea, we constructed a method for assessing the grade of diabetic retinopathy based on a threshold inspection mechanism. The overall process of the method is shown in Fig. 5. As a whole, the method can be roughly divided into the following four steps:

In the *first* step (classification algorithm training), three data augmentation methods (MixUp [27], CutMix [25], and Crop) are used to avoid the EfficientNet algorithm [20] from falling into the problem of overfitting.

In the *second* step, we obtain the pretrained weights of the EfficientNet algorithm [20] and perform DR grade assessment on the test images. After this step is completed, we can get the preliminary results of DR grade classification.

In the *third* step, according to the weights of MCS-DRNet, we construct a grade revision mechanism based on pixel thresholds in the lesion area. The role of this mechanism is to adjust the results of the preliminary classification in the second step. In this mechanism, the MCS-DRNet method is responsible for identifying various lesion areas in the image, and accumulating and outputting the pixel values of these areas. One thing needs to be explained in particular, the MCS-DRNet method deployed in this framework is different from the MCS-DRNet v1 method. The main difference is that some of the output items are

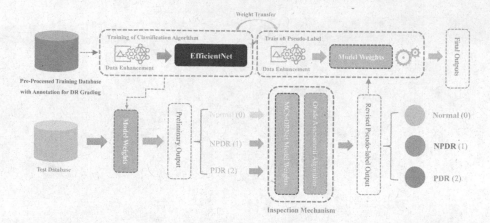

Fig. 5. Evaluation method of diabetic retinopathy grade based on threshold inspection mechanism.

different. Equation 4 lists the output terms of the MCS-DRNet method adopted in this framework.

$$
\begin{cases}
O_1 = O_1^{\varphi_{1536R}^m} \cup O_1^{\varphi_{1024}^s} \cup O_1^{\varphi_{1536R}^s} \\
O_2 = O_2^{\varphi_{1536R}^m} \\
O_3 = O_3^{\varphi_{1536R}^m} \cup O_3^{\varphi_{1024}^s} \cup O_3^{\varphi_{1536R}^s}
\end{cases}
\tag{4}
$$

Table 1. T_i (and T_i^*) is used to judge the pixel threshold of output O_i, where i equals $(1, 2, 3)$.

Threshold	T_1	T_2	T_3	T_1^*	T_2^*	T_3^*
Value	26_2	130_2	28_2	78_2	750_2	100_2

Based on Eq. 4, we construct the inspection and revision mechanism as follows:

First, Eq. 5 shows the sentence used to judge the pixel threshold condition judgment of each output item (O_1, O_2, O_3).

$$
\begin{cases}
C_i^{\min} : O_i < T_i \\
C_i^{\max} : O_i > T_i^*
\end{cases}
\tag{5}
$$

The conditional statements C_i^{min} and C_i^{max} are mainly used for the diagnosis of disease severity, which correspond to Normal and PDR respectively. In Eq. 5, the values of i are $(1, 2, 3)$, Table 1 gives the thresholds used for pixel judgment in each image.

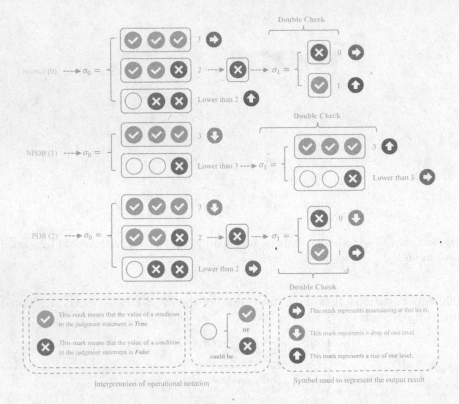

Fig. 6. The specific principle of the threshold inspection mechanism. The checks and adjustments for Normal (0), NPDR (1) and PDR (2) are in parallel, that is, the results of the adjustments do not affect each other. There is no ordering relationship between the elements (correct and incorrect symbols) within each parenthesis. All adjustments will only be made between adjacent levels.

Secondly, Eq. 6 shows the cumulative calculation formula of the judgment conditions true and false.

$$f_c(x) = \begin{cases} \text{if } x \text{ equals True, return } 1 \\ \text{if } x \text{ equals False, return } 0 \end{cases} \tag{6}$$

Finally, Eqs. 7 and 8 give the overall calculation formula of the pixel threshold inspection mechanism.

$$\sigma_0 = \sum_{i=1}^{3} f_c\left(C_i^{\min}\right) \tag{7}$$

$$\sigma_1 = \sum_{i=1}^{3} f_c\left(C_i^{\max}\right) \tag{8}$$

Based on Eqs. 7 and 8, we constructed the diabetic retinopathy grade inspection and adjustment mechanism, as shown in Fig. 6.

For the Normal (0) level, our adjustment scheme is as follows:

(1) $\sigma_0 = 3$, which means that the sample satisfies all the conditions for the level of Normal, so it remains unchanged at this level;
(2) $\sigma_0 = 2$, which means that the sample only satisfies some of the conditions for the Normal level. We need further confirmation of that *False* condition. This False condition can be identified as NPDR as long as it satisfies any of the conditions in the σ_1 judgment statement, instead, it remains unchanged at that level
(3) $\sigma_0 \leq 1$, which means that the sample is not enough to be recognized as Normal. Therefore, the sample needs to be upgraded to NPDR.

For the NPDR (1) level, our adjustment scheme is as follows:

(1) $\sigma_0 = 3$, which means that the sample satisfies all conditions for level Normal. Therefore, this sample is downgraded to Normal.
(2) $\sigma_0 \leq 2$, which means that it is reasonable for this sample to be considered NPDR, but further confirmation is required. If the sample meets all the conditions of σ_1, it is deemed unreasonable at the NPDR level and will be upgraded to PDR, otherwise, it will remain at that level.

For the PDR (2) level, our adjustment scheme is as follows:

(1) $\sigma_0 = 3$, which means that the sample does not meet the level of PDR and needs to be lowered by one level.
(2) $\sigma_0 = 2$, which means that the sample may not meet the level of PDR, We need further confirmation of that *False* condition. As long as this *False* condition, in the judgment statement σ_1, satisfies a condition, it can be considered that the sample is divided into PDR is reasonable, otherwise, it is unreasonable
(3) $\sigma_0 \leq 1$, which means that the sample is reasonable at the level of PDR, so it will be retained at this level.

In the *last* step (step 4), the adjusted pseudo-annotations are used to train EfficientNet v2, and the final result is output. In this step, the methods of data enhancement are MixUp [27], and CutMix [25].

3 Experimental Settings and Results

3.1 Segmentation of Diabetic Retinopathy Tissue

In the experimental framework for segmentation of diabetic retinopathy tissue, first, we do semi-supervised pre-training in DRAC2022 DR grading assessment training set. Then, on the training set of DRAC2022 semantic segmentation, we perform supervised training to evaluate the representation by end-to-end fine-tuning. In addition, the results obtained by various experimental methods on the segmentation leaderboard are presented. Finally, to further verify the effectiveness of the proposed method, we visualize the semantic segmentation results of the three DR categories (intraretinal microvascular abnormals, nonperfusion areas, and neovascularization) on the test set.

Table 2. Data augmentation results of DRAC2022 DR grading dataset. 0 to 2 represent the three DR levels of Normal, NPDR, and PDR in turn.

Level of assessment	0	1	2	Total
Amount of raw data	328	213	70	611
Quantity after data augmentation	1968	1278	420	3666

Table 3. Data augmentation results of the DRAC2022 semantic segmentation dataset. The data is enhanced by flipping vertically and horizontally and rotating counterclockwise 90 °C, 180 °C, 270 °C. Since nonperfusion areas in the image has more pixel values than classes intraretinal microvascular abnormalities and neovascularization. In order to avoid the class unbalance problem that may be caused by the algorithm in the identification process of each category, we write the ground truth of nonperfusion areas to Mask B, and the ground truth of the remaining two lesion types is written on Mask A.

Types of mask		A	B
The original data in the data set	Number of images with lesions present	88	106
	Number of images without lesions	21	3
Quantity after data augmentation	Number of images with lesions present	528	636
	Number of images without lesions	126	18

Semi-Supervised Pre-training (Classification Algorithm Training). In the semi-supervised pre-training part of MAE [10], to make the pre-training algorithm more fully mine the supervised information contained in UW-OCTA images, so as to improve the performance of the deep learning algorithm in processing semantic segmentation task. We use the DR assessment section of the DRAC2022 dataset. The dataset contains a total of 611 original images. We process these images by rotating 90 °C, 180 °C, and 270 °C as well as flipping horizontally and vertically to acquire a total of 3666 images for pre-training, as shown in Table 2. In addition, we do various data enhancements during pre-training, including brightness, contrast, saturation and hue.

The training parameters of MAE algorithms are as follows: backbone (vit-base); resized crop (480); base learning rate (1.5e–4); warm-up iteration (40); max epochs (1600).

Supervised Learning (Segmentation Algorithm Training). The original data of the semantic segmentation dataset provided by the DRAC2022 challenge has 109 images with a pixel value of 1024×1024. It includes the following three diagnostic categories: intraretinal microvascular abnormals, nonperfusion areas, and neovascularization. In the image data of these three categories, there are 86, 106, and 35 images with manual annotation (lesions in the images), respectively. In the process of the experiment, to avoid the algorithm falling into over-fitting and improve the generalization of the algorithm, we augment 109 images in all the original data, and the augmentation results are shown in Table 3. The image data of each category are angle-adjusted and expanded to a total of 654 images (including the original image). We believe that in the training process, adding images of non-diseased areas (background images) is conducive to enhancing the learning of the deep learning algorithm for non-diseased areas and reducing the probability of misjudgment of the algorithm for diseased areas. In addition, we also do various data enhancements during the training, including brightness, contrast, saturation and hue.

In algorithm training, we train three algorithms, which are pre-trained MAE, SegFormer, and ConvNeXt. The training parameters of these algorithms are as follows:

(1) The parameters of pre-trained MAE algorithm training are as follows: pre-training algorithm (vit-base); crop size (640^2); base learning rate (1e–6); optimizer (AdamW); max iterations (130k); loss (CrossEntropyLoss).
(2) The parameters of SegFormer algorithm training are as follows: pre-training algorithm (mit-b3); crop size (1024^2); base learning rate (1e–6); optimizer (AdamW); max iterations (130k); loss (CrossEntropyLoss).
(3) The parameters of ConvNeXt algorithm training are as follows: pre-training algorithm (ConvNeXt L/XL); crop size (640^2); base learning rate 8e-6; optimizer (AdamW); max iterations (110k for L / 130k for XL); loss (CrossEntropyLoss).

Experimental Results. For quantitative evaluation, Table 4 shows the mean Intersection over Union (mIoU) and mean Dice (mDice) of subclass 1 and 3 for algorithms UNet + DeepLabV3 [16] and MCS-DRNet series methods in the DRAC2022 semantic segmentation data set. As you can see from the table, the MCS-DRNet series of methods constructed in this paper significantly outperformed the traditional medical semantic segmentation algorithm UNet + DeepLabV3 [16] in identifying the mIoU and mDice of two subcategories (intraretinal microvascular abnormals and neovascularization).

Among the MCS-DRNet family of algorithms, the original MCS-DRNet v2 (A.3) method achieves 5.26% and 0.39% improvement on the mDice metric for categories 1 and 3, respectively, compared to v1. Moreover, in the original MCS-DRNet v2 method, when we add multi-scale (MS) test image operation (A.4), the mDice measurement index of Classes 1 and 3 is improved by 0.1% and 0.44% respectively. This means that this operation can effectively improve the performance of our method in the identification of DR areas at multiple scales. Based on performing the MS, we added the multi-angle rotation (MR) operation to obtain our final MCS-DRNet v2 method (A.5). The addition of this method increased the mDice index of Classes 1 and 3 by 2.87% and 0.16%, respectively.

Table 5 presents the mIoU and mDice of UNet+DeepLabV3 and MCS-DRNet series methods in identifying Class 2 in the DRAC2022 semantic segmentation dataset. As can be seen from the table, MCS-DRNet v2 based on ConvNeXt-XL (B.4) achieves the best results in both mIoU and mDice metrics in the task of identifying Class 2. However, when we conduct MR operations on the MCS-DRNet v2 method (B.3), the measured values of mIoU and mDice both decreased to varying degrees.

Table 4. Comparison of the performance of MCS-DRNet in identifying intraretinal microvascular abnormalities (Class 1) and neovascularization (Class 3) of lesion classes. In the parameter Image Size, M represents the existence of two resolutions (640×640 and 1024×1024) in this method. In the parameter Backbone, the letters M, C, and S indicate the integration of MAE (ViT-base) [10], ConvNeXt-XL [13] and SegFormer-B3 [24], respectively. *w/o* is short for *without*.

ID	Method	Backbone	Image Size	Mask A			
				Class 1		Class 3	
				mIoU(%)	mDice(%)	mIoU(%)	mDice(%)
A.1	UNet + DeepLabV3 [16]	UNet-S5-D16	1024×1024	13.85	22.35	33.80	48.39
A.2	MCS-DRNet v1	MS	M	22.83	35.79	41.06	57.04
A.3	MCS-DRNet v2 (*w/o* MR & MS)	MCS	M	26.98	41.05	41.56	57.43
A.4	MCS-DRNet v2 (*w/o* MR)	MCS	M	27.30	41.15	41.86	57.87
A.5	MCS-DRNet v2	MCS	M	**29.63**	**44.02**	**42.13**	**58.03**

Table 5. Performance comparison of various algorithms in identifying nonperfusion areas (Class 2/ Mask B). *w* is short for *with*.

ID	Method	Backbone	Image Size	mIoU(%)	mDice(%)
B.1	UNet + DeepLabV3 [16]	UNet-S5-D16	1024×1024	41.06	55.50
B.2	MCS-DRNet v1	ConvNeXt-L	640×640	47.78	61.98
B.3	MCS-DRNet v2 (*w* MR)	ConvNeXt-XL	640×640	48.74	62.96
B.4	MCS-DRNet v2	ConvNeXt-XL	640×640	**49.99**	**64.26**

Finally, Table 6 shows the final mean dice similarity coefficient (mean DSC) obtained by all methods. It can be seen that the MCS-DRNet v2 (composed of A.5 and B.4) outperforms other methods.

In order to further verify the effectiveness of MCS-DRNet series methods in this paper. Figure 7 (a)–(c) sequentially show the recognition effects of the proposed method on three DR lesion categories (intraretinal microvascular abnormalities, nonperfusion areas and neovascularization) on the test set of DRAC2022. From the images, it can be seen that MCS-DRNet can well identify the regions of various DR areas in UW-OCTA images.

In order to more intuitively show the various algorithms in the proposed method, and the role play in these identification results. Figure 8 and Fig. 9 demonstrate the recognition effect of these sub-algorithms on intraretinal microvascular abnormals and neovascularization for DR categories, respectively. In these two images, we annotate the unique contribution (patches 1–4) of each sub-algorithm in the recognition of pathological tissue features in the whole image.

From Fig. 8, it is not difficult to find that the MAE algorithm based on the self-attention structure can well excavate some unique subtle DR tissue. The ConvNeXt algorithm based on the "sliding window vision" strategy can find DR

Table 6. Comparison of Mean DSC for each model and algorithm. The *asterisk* (*) in the upper right corner of the number represents the final result of the DRAC2022 challenge. The following table follows this principle.

ID	Backbone (Mask A & B)	mean DSC (%)
A.1 & B.1	UNet-S5-D16	42.08
A.2 & B.2	MS & ConvNeXt-L	51.61*
A.3 & B.4	MCS & ConvNeXt-XL	54.23
A.4 & B.4	MCS & ConvNeXt-XL	54.41
A.5 & B.4	MCS & ConvNeXt-XL	**55.44**

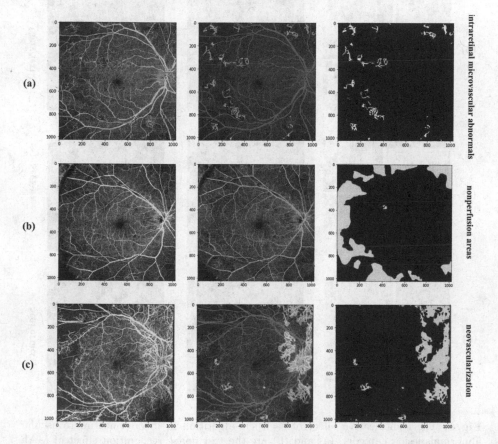

Fig. 7. The recognition effect of the method in this paper in three sub-categories of lesion areas.

tissue in the image more evenly. The segFormer algorithm plays an important role in the recognition of lesions of different scales in images.

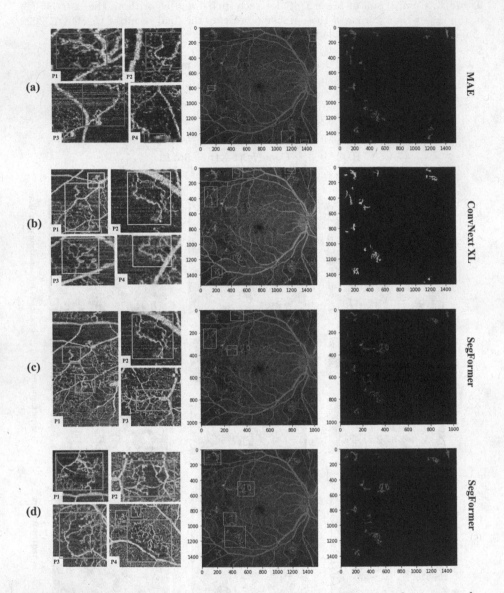

Fig. 8. The identification effect of each sub-algorithm in intraretinal microvascular abnormals lesion category. (a) and (b) are the test image recognition effect of MAE and ConvNeXt at a resolution of 1536 × 1536, respectively. (c) and (d) represent the recognition effect of SegFormer on the scale of 1024×1024 and 1536×1536, respectively. P1- P4 represent patches 1–4 in turn. P1 - P4 in the leftmost column corresponds to the enlargement of P1 - P4 in the middle column. The groundtruth with different colors in each Patch in the leftmost column represents the area where the diseased tissue is located. Figure 9 also follows these principles.

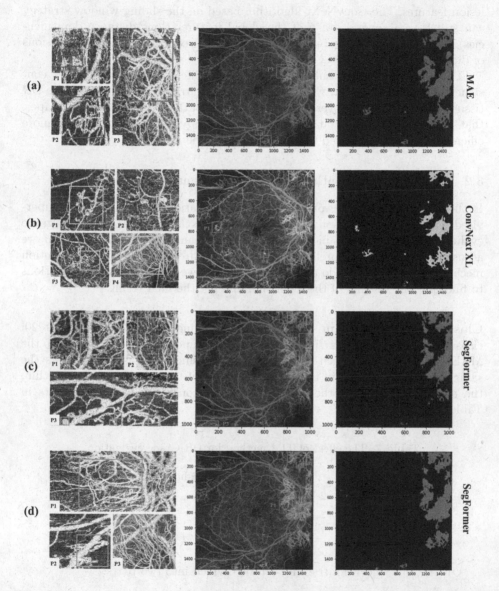

Fig. 9. Recognition effect of each sub-algorithm in neovascularization lesion category. This diagram follows the same principles as Fig. 8.

From Fig. 9, we can see that the MAE algorithm and SegFormer based on the self-attention architecture can pay attention to the most obvious areas in the image, and at the same time, they can also pay attention to some subtle lesion features. The ConvNeXt algorithm based on the sliding window strategy can still have a general grasp of the global features (long-distance dependencies). SegFormer is very robust in the recognition of neovascularization lesions of different sizes.

Therefore, the set of sub-algorithms with different visual strategies constructed in this paper have their own unique contributions in identifying DR tissue. This result further verifies the rationality of the hypothesis in this paper, that is, our method integrating different visual processing strategies can more efficiently mine the diseased tissue contained in the image.

3.2   Diabetic Retinopathy Grade Assessment

In the experimental framework of DR grade evaluation, first, we perform supervised learning (classification algorithm training) on the DRAC2022 DR grade evaluation data set to obtain the results of the preliminary evaluation. Then, we adjust the results of the preliminary evaluation through a threshold inspection mechanism to obtain pseudo-annotations. Finally, we use the pseudo-annotations to fine-tune the weights of the algorithm and get the final result.

Classification Algorithm Training. The training of the EfficientNet v2 [20] classification algorithm is similar to the semi-supervised pre-training of the MAE [10] algorithm, which is essentially a classification task and shares the same dataset. Therefore, in the training process of the classification algorithm, the expansion results of the training data are consistent with the results in Table 2.

Table 7. The effect of TIM method for DR Grade evaluation.

Method	Quadratic weighted kappa (%)
EfficientNet v2 [20]	72.60
EfficientNet v2 + TIM	**75.59***

Overall, our algorithm training can be divided into two parts.

(1) The first part is the training based on the DR Level assessment data set. The training parameters of this part are as follows: experimental algorithm (EfficientNet v2 M); initialization of weights (ImageNet); max epochs (220); learning rate warmup epoch (40); dropout (0.5); data enhancement (Mixup, CutMix, and Crop).

(2) The second part is the training based on the pseudo-labeled dataset. The training parameters of this part are as follows: experimental algorithm (Efficientnet v2 M); initialization of weights (the weight of the first part of the algorithm); max epochs (100); learning rate warmup epoch (10); dropout (0.2); data enhancement (Mixup and Cutmix).

The classification algorithm training image size of these two parts is 480^2.

Experimental Results and Analysis. For quantitative evaluation, Table 7 lists the results obtained by EfficientNet v2 [20] and EfficientNet v2 + threshold inspection mechanism (TIM) on the competition list. As can be seen from the table, after using the TIM method, the EfficientNet v2 brings a 2.99% improvement in quadratic weighted kappa. To some extent, the results verify the rationality of our conjecture about the difference between classification and segmentation algorithms in obtaining UW-OCTA image features.

Finally, we summarize three shortcomings of the framework. First of all, DR grading assessment method based on threshold value, the selection of threshold value is highly professional. If the threshold selection is good, the performance of the algorithm to identify the DR Organization could be greatly improved. Second, in this challenge, due to the urgency of time, the semantic segmentation model we built for grade evaluation is relatively simple and needs to be improved. Last but not least, in this challenge, the method we submitted in the system only performed one iteration between the *third* step and the *fourth* step. However, multiple iterations can be beneficial for more optimal results.

4   Conclusion

Typically, microvascular hemorrhage results in the formation of new blood vessels that subsequently transition to proliferative DR. This is potentially harmful to vision. To facilitate the development of diabetic retinopathy automatic detection, this paper proposes a novel semi-supervised semantic segmentation method for UW-OCTA DR grade assessment. In this method, firstly, supervised DR feature information in UW-OCTA images is mined through semi-supervised pre-training. Secondly, a cross-algorithm integrated DR Semantic segmentation algorithm is constructed according to these characteristics. Finally, a DR grading evaluation method is constructed based on this semantic segmentation algorithm. The experimental results verify that our method can accurately segment most of the images classified as proliferative DR and assess the DR Level reasonably.

References

1. Bao, H., Dong, L., Wei, F.: Beit: bert pre-training of image transformers. arXiv preprint arXiv:2106.08254 (2021)
2. Caron, M., Bojanowski, P., Mairal, J., Joulin, A.: Unsupervised pre-training of image features on non-curated data. In: 2019 IEEE/CVF International Conference on Computer Vision (ICCV), pp. 2959–2968 (2019)

3. Caron, M., Touvron, H., Misra, I., Jégou, H., Mairal, J., Bojanowski, P., Joulin, A.: Emerging properties in self-supervised vision transformers. In: Proceedings of the IEEE/CVF International Conference on Computer Vision, pp. 9650–9660 (2021)
4. Chen, X., He, K.: Exploring simple siamese representation learning. In: Proceedings of the IEEE/CVF Conference on Computer Vision and Pattern Recognition, pp. 15750–15758 (2021)
5. Contributors, M.: MMSegmentation: Openmmlab semantic segmentation toolbox and benchmark. https://github.com/open-mmlab/mmsegmentation (2020)
6. Contributors, M.: MMSelfSup: Openmmlab self-supervised learning toolbox and benchmark. https://github.com/open-mmlab/mmselfsup (2021)
7. Dai, L., et al.: A deep learning system for detecting diabetic retinopathy across the disease spectrum. Nat. Commun. **12**(1), 1–11 (2021)
8. Doersch, C., Gupta, A.K., Efros, A.A.: Unsupervised visual representation learning by context prediction. In: 2015 IEEE International Conference on Computer Vision (ICCV), pp. 1422–1430 (2015)
9. DosoViTskiy, A., et al.: An image is worth 16x16 words: transformers for image recognition at scale. arXiv preprint arXiv:2010.11929 (2020)
10. He, K., Chen, X., Xie, S., Li, Y., Dollár, P., Girshick, R.: Masked autoencoders are scalable vision learners. arXiv preprint arXiv:2111.06377 (2021)
11. Khan, A., AlBarri, S., Manzoor, M.A.: Contrastive self-supervised learning: a survey on different architectures. In: 2022 2nd International Conference on Artificial Intelligence (ICAI), pp. 1–6. IEEE (2022)
12. Liu, R., et al.: Deepdrid: diabetic retinopathy-grading and image quality estimation challenge. Patterns **3**(6), 100512 (2022)
13. Liu, Z., Mao, H., Wu, C.Y., Feichtenhofer, C., Darrell, T., Xie, S.: A convnet for the 2020s. In: Proceedings of the IEEE/CVF Conference on Computer Vision and Pattern Recognition (CVPR) (2022)
14. Madzin, H., Zainuddin, R.: Feature extraction and image matching of 3d lung cancer cell image. In: 2009 International Conference of Soft Computing and Pattern Recognition, pp. 511–515. IEEE (2009)
15. Madzin, H., Zainuddin, R., Mohamed, N.S.: Analysis of visual features in local descriptor for multi-modality medical image. Int. Arab J. Inf. Technol. (IAJIT) **11**(5) (2014)
16. Ronneberger, O., Fischer, P., Brox, T.: U-net: convolutional networks for biomedical image segmentation. In: Navab, N., Hornegger, J., Wells, W.M., Frangi, A.F. (eds.) MICCAI 2015. LNCS, vol. 9351, pp. 234–241. Springer, Cham (2015). https://doi.org/10.1007/978-3-319-24574-4_28
17. Sheng, B., et al.: An overview of artificial intelligence in diabetic retinopathy and other ocular diseases. Front. Public Health **10**, 971943 (2022)
18. Sheng, B., et al.: Diabetic retinopathy analysis challenge 2022, March 2022. https://doi.org/10.5281/zenodo.6362349
19. Srivastava, N.: Unsupervised learning of visual representations using videos (2015)
20. Tan, M., Le, Q.: Efficientnetv2: smaller models and faster training. In: International Conference on Machine Learning, pp. 10096–10106. PMLR (2021)
21. Tan, Z., Hu, Y., Luo, D., Hu, M., Liu, K.: The clothing image classification algorithm based on the improved xception model. Int. J. Comput. Sci. Eng. **23**(3), 214–223 (2020). https://doi.org/10.1504/ijcse.2020.111426
22. Wang, X., He, K., Gupta, A.K.: Transitive invariance for self-supervised visual representation learning. In: 2017 IEEE International Conference on Computer Vision (ICCV), pp. 1338–1347 (2017)

23. Wu, Z., Xiong, Y., Yu, S.X., Lin, D.: Unsupervised feature learning via non-parametric instance discrimination. In: 2018 IEEE/CVF Conference on Computer Vision and Pattern Recognition, pp. 3733–3742 (2018)
24. Xie, E., Wang, W., Yu, Z., Anandkumar, A., Alvarez, J.M., Luo, P.: Segformer: simple and efficient design for semantic segmentation with transformers. arXiv preprint arXiv:2105.15203 (2021)
25. Yun, S., Han, D., Oh, S.J., Chun, S., Choe, J., Yoo, Y.: Cutmix: regularization strategy to train strong classifiers with localizable features. In: Proceedings of the IEEE/CVF International Conference on Computer Vision, pp. 6023–6032 (2019)
26. Zhai, X., Oliver, A., Kolesnikov, A., Beyer, L.: S4l: self-supervised semi-supervised learning. In: Proceedings of the IEEE/CVF International Conference on Computer Vision, pp. 1476–1485 (2019)
27. Zhang, H., Cisse, M., Dauphin, Y.N., Lopez-Paz, D.: mixup: beyond empirical risk minimization. arXiv preprint arXiv:1710.09412 (2017)

Image Quality Assessment Based on Multi-model Ensemble Class-Imbalance Repair Algorithm for Diabetic Retinopathy UW-OCTA Images

Zhuoyi Tan[ID], Hizmawati Madzin[✉][ID], and Zeyu Ding[ID]

Faculty of Computer Science and Information Technology, Universiti Putra Malaysia, Serdang 43400, Malaysia
hizmawati@upm.edu.my

Abstract. In the diagnosis of diabetic retinopathy (DR), ultrawide optical coherence tomography angiography (UW-OCTA) has received extensive attention because it can non-invasively detect the changes of neovascularization in diabetic retinopathy images. However, in clinical application, UW-OCTA digital images will always suffer a variety of distortions due to a variety of uncontrollable factors, and then affect the diagnostic effect of DR. Therefore, screening images with better imaging quality is very crucial to improve the diagnostic efficiency of DR. In this paper, to promote the development of UW-OCTA DR image quality automatic assessment, we propose a multi-model ensemble class-imbalance repair (MMECIR) algorithm for UW-OCTA DR image quality grading assessment. The models integrated with this algorithm are ConvNeXt, EfficientNet v2, and Xception. The experimental results show that the MMECIR algorithm constructed in this paper can be well applied to UW-OCTA diabetic retinopathy image quality grading assessment (the quadratic weighted kappa of this algorithm is 0.6578). Our code is available at https://github.com/SupCodeTech/DRAC2022.

Keywords: Image quality assessment · UW-OCTA image · Deep learning

1 Introduction

The main advantage of ultra-wide optical coherence tomography (UW-OCTA) compared to the traditional hierarchical diagnostic methods of diabetic retinopathy (DR) (fundus photography and FFA) is that the tiny blood vessels in the retina and choroid can be clearly seen. This advantage makes UW-OCTA a pivotal imaging method for ophthalmologists to diagnose PDR today [3,4,8,9].

Supported by Ministry of Higher Education, Malaysia.

B. Sheng and M. Aubreville (Eds.): MIDOG 2022/DRAC 2022, LNCS 13597, pp. 118–126, 2023.
https://doi.org/10.1007/978-3-031-33658-4_11

However, in clinical applications, UW-OCTA digital images may experience various distortions during acquisition, processing, compression, storage, transmission, and reproduction, any of which may lead to a decline in visual quality [13], which in turn affects the diagnostic efficacy of DR. Therefore, screening out high-quality UW-OCTA images with better imaging is crucial to improve the diagnostic efficiency of DR.

In this paper, to promote the development of UW-OCTA DR image quality automatic assessment, we propose a multi-model ensemble class-imbalance repair (MMECIR) algorithm for UW-OCTA DR image quality grading assessment [1,2,5,11,12]. In the implementation process of this algorithm, firstly, we generate many low-quality category images (through high-quality images corruption) in the data preprocessing stage, to alleviate the problem of an unbalanced distribution of image data of the three evaluation levels in the data set (low to medium-quality images have far less data than high-quality images). Secondly, to avoid the bias caused by a single model structure to the result prediction and improve the credibility of the prediction result. We train three models, ConvNeXt [5], EfficientNet v2 [11], and Xception [1], simultaneously to obtain preliminary prediction results of multi-model ensembles. We then feed the results of this preliminary prediction into an image quality assessment algorithm to obtain pseudo-annotations. Finally, we perform the final training on the pseudo-labels and output the challenge results.

On the diabetic retinopathy analysis challenge (DRAC) 2022 test set, the baseline method (challenge version) and the MMECIR algorithm (post-challenge version) achieved 0.6238 and 0.6578 quadratic weighted kappa results in the DR image quality grade evaluation, respectively. This result verifies that our method can reasonably evaluate the DR image quality level.

2   Approach

2.1   Baseline Method (Challenge Version)

As a whole, the baseline method (single model method) in this paper can be divided into two stages:

The task performed in the first stage is data pre-processing. In this stage, firstly, we uniformly perform image orientation adjustment (OA) (horizontal and vertical flip, counterclockwise rotation 90, 180, 270) operations on all data of the three image quality evaluation levels. Secondly, on the basis of the image OA operation, we generate a large number of low-quality images by performing image corruption operations on the pre-processed high-quality images, so as to alleviate the imbalance of the category data between the low-quality and high-quality images. The specific implementation principle of the image corruption operation can be divided into the following two steps:

In the first step, a complete image is divided into 16 image patches (the implementation principle is shown in Eq. 1).

$$\sum_{i=1}^{16} \overleftarrow{P_i} = f_D(x) \tag{1}$$

Among them, x represents a complete input image. Function f_D is responsible for dividing the whole image x into 16 patches. $\overleftarrow{P}_i$ $(i = 1, ..., 16)$ represents a certain area that is divided in a complete image x (the arrow to the left indicates that the patch contains position information).

In the second step, we randomly select λ patches to perform the mask operation (the implementation principle is shown in Eq. 2). Where λ coefficient mainly plays two roles: First, it is to avoid the model from forming a memory of the area size of the damaged area in the image. Second, to better simulate the real situation of image damage. The coefficient ranges from 3 to 8. The implementation effect of our image corruption algorithm ($\lambda = 8$) is shown in Fig. 1.

$$y = M_{16}^{\lambda} \odot \sum_{i=1}^{16} \overleftarrow{P}_i \tag{2}$$

Among them, M_{16}^{λ} means to generate a mask image with λ patches masked randomly. The operator $\odot$ represents the pixel value of the masked area is set to 0. y represents the masked training image.

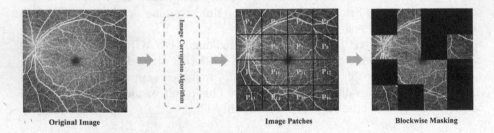

Fig. 1. Implementation process of image corruption algorithm.

The task performed in the second stage is the algorithm training and the output of the final result. In this stage, first, we train the ConvNeXt [5] on the pre-processed dataset. Then, predict the test set according to the training weights of the ConvNeXt [5], and get the final prediction result.

2.2 MMECIR Algorithm (Post-challenge Version)

The DR image quality assessment method based on the MMECIR algorithm can be divided into four stages as a whole (as shown in Fig. 2):

The task performed in the first stage is data pre-processing. In this stage, the data pre-processing method is the same as the baseline method, so it will not be repeated here.

The tasks performed in the second stage are the training of the algorithm and the pre-evaluation of the test set. In this stage, first, we train three models

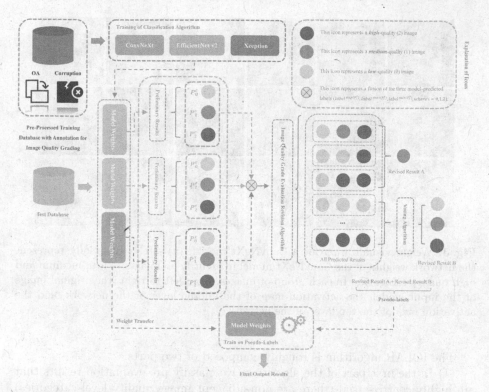

Fig. 2. The implementation principle of MMECIR algorithm. OA stands for orientation adjustment.

ConvNeXt [5], EfficientNet v2 [11], and Xception [1], simultaneously on the preprocessed dataset. Then, predict the test set according to the training weights of these three models, and get preliminary prediction results.

The tasks performed in the third stage is to revise the results of the preassessment. In this stage, first of all, in order to avoid the deviation caused by the single model structure on the result prediction and improve the reliability of the prediction results, we use the pre-evaluation results as the *input* of the image quality level assessment revision (IQLAR) algorithm. The mathematical expression of this *input* is shown in Eq. 3.

$$\text{input} = \left(\text{label}^{\max(P_i^c)}, \text{label}^{\max(P_i^e)}, \text{label}^{\max(P_i^x)} \right) \tag{3}$$

Among them, $\text{label}^{\max(P_i^\varrho)}$ represents the maximum probability label predicted by model ϱ ($\varrho = c, e, x$). There is no order between $\text{label}^{\max(P_i^c)}$, $\text{label}^{\max(P_i^e)}$ and $\text{label}^{\max(P_i^x)}$. The value of i is $0, 1, 2$, which represent low, medium and high quality class image, respectively.

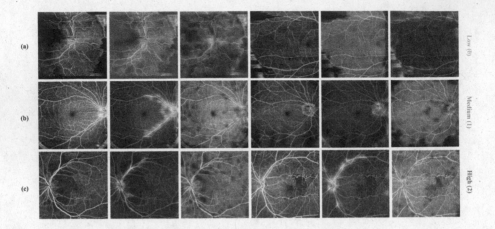

Fig. 3. Weight visualization for a ConvNeXt model. (a) - (c) sequentially represent the network weights of the ConvNeXt model in identifying images of low, medium and high quality categories. In each group of images, from left to right, the original image of the input model, the activation map of the middle layer of the network, and the activation map of the top layer of the network.

The IQLAR algorithm is roughly composed of two parts:

(1) In the first part of the algorithm, we classify pre-evaluation results that are highly controversial (there are non-adjacent image quality level categories) into images of medium quality category (revised result A). The process is shown in Eq. 4.

$$\text{revised label } A = f_{na}(\text{ input}) \tag{4}$$

Among them, the function f_{na} is responsible for determining whether there are non-adjacent categories. The implementation principle of this function is shown in Eq. 5.

$$f_{na}(\alpha, \beta, \gamma) = \begin{cases} \text{if } |\alpha - \beta| > 1, \text{ return } 1 \\ \text{if } |\alpha - \gamma| > 1, \text{ return } 1 \\ \text{if } |\beta - \gamma| > 1, \text{ return } 1 \end{cases} \tag{5}$$

The reason for this is twofold. First, it is to alleviate the problem of low data quantity distribution in the pre-evaluation results for the medium quality category images. Second, in our opinion, for an image, there is a high degree of controversy in the pre-assessment results, which means that the image is likely to have some unique features of both high quality and low quality images. Therefore, it is logical to classify these highly controversial pre-assessment results into medium quality images.

(2) In the second part of the algorithm, we feed the results of the pre-evaluation that are less controversial (categories with only adjacent image quality levels) into the voting algorithm, resulting in revised result B. The basic principle of the voting algorithm is: the category with the highest frequency in the

Table 1. Data augmentation results of DRAC2022 DR image quality grading dataset. 0 to 2 represent the three image quality levels of low, medium, and high in turn.

Level of assessment	0	1	2	Total
Amount of raw data	50	97	518	665
Quantity after image orientation adjustment	300	582	3108	3990
Quantity after image corruption algorithm	3408	582	3108	7098

input is used as the result of algorithm revision. The principle of this process is shown in Eq. 6.

$$\text{revised label} \quad B = f_a \,(\text{input}) \tag{6}$$

where, the function completed by f_a is to return one of the two categories as long as there are two identical categories in the collection *input*. The mathematical expression of the function f_a is shown in Eq. 7.

$$f_a(\alpha, \beta, \gamma) = \begin{cases} \text{if } \alpha = \beta, & \text{return } \alpha \\ \text{if } \alpha = \gamma, & \text{return } \alpha \\ \text{if } \beta = \gamma, & \text{return } \beta \end{cases} \tag{7}$$

The task performed in the fourth stage is the training based on the pseudo-labeled data set and the output of the final result. In this stage, first, we merge revised result A and revised result B into pseudo-labels and form a pseudo-labeled dataset. Then, the pre-trained ConvNeXt is fine-tuned on this pseudo-labeled dataset, and the final result is output. Finally, we visualize the model weights of ConvNeXt in recognizing low-medium-high-quality category images, as shown in Fig. 3.

3   Dataset

The DR image quality assessment dataset for the DRAC2022 Challenge contains a total of 665 images. Each image has a pixel value size of 1024×1024. In the data pre-processing stage, firstly, we carried out orientation-adjustment for all kinds of images. By rotating 90 °C, 180 °C and 270 °C, and horizontal and vertical rotation, 300, 582 and 3108 images of low, medium and high quality were obtained respectively. Secondly, the image corruption algorithm was used to process the 3108 high-quality images and generate 3108 low-quality images to alleviate the problem of data imbalance between high and low quality categories. Table 1 shows the result of data pre-processing.

4   Implementation

The training of the MMECIR algorithm can be divided into two parts: the first part is the training based on DR image quality grade assessment data set,

In this part, the experimental models we use are ConvNeXt-L, EfficientNet v2 (M), and Xception. The maximum training epochs for these three models are 125, 150, and 160, respectively. The initialized weights are all ImageNet. The second part is the training based on the pseudo-labeled dataset. In this part, we train ConvNeXt-L on the pseudo-labeled dataset and output the final result. The model has a maximum training epoch of 80. The initialization method is the training weight of the first part.

In addition, during the training process of the above two parts, we abandon the MixUp [15], CutMix [14] and Dropout [10] methods (the dropout method removes information pixels on the training image by superimposing black pixels or random noise, which will affect the effect of model learning to some extent) to avoid disturbance of the training image quality.

5 Evaluation and Results

For quantitative evaluation, Table 2 shows the performance of our method on the image quality assessment test set in the DRAC2022 challenge. As can be seen from the table, the performance of the MMECIR algorithm in image quality assessment is higher than that of the baseline method based on a single model, reaching 65.78% quadratic weighted kappa. Moreover, the image erosion operation brings a 2.33% increase in quadratic weighted kappa to our baseline method. The experimental results verify the rationality of the two hypotheses in this paper:

(1) This paper builds a data class balance repair algorithm (the algorithm is composed of an image quality corruption algorithm and an image quality level revision algorithm), which is conducive to alleviating the impact of each category imbalance on the recognition performance.
(2) The image quality assessment method based on multi-model integration can effectively overcome the bias caused by a single model structure in recognition.

Table 2. Experimental results of DRAC2022 image quality test set. *w/o* is short for without. The asterisk (*) in the upper right corner of the number represents the final result of the DRAC2022 challenge.

Method	Quadratic weighted kappa (%)
Baseline method (*w/o* corruption)	60.05
Baseline method	62.38*
MMECIR Algorithm	**65.78**

6   Conclusion

In clinical applications, UW-OCTA images are always inevitably damaged, which will have a bad impact on the diagnosis of DR. Therefore, in order to promote the development of automatic assessment of UW-OCTA image quality for diabetic retinopathy, this paper constructs a UW-OCTA image quality grade assessment method based on multi-model ensemble class-imbalance repair algorithm. In this method, first, in the data preprocessing stage, a large number of images of low-quality categories are generated through an image quality corruption algorithm, so that the distribution of image data of high-quality and low-quality categories tends to be balanced. Second, a preliminary evaluation of the image quality level is carried out by means of cross-models integration. Thirdly, the pre-evaluation results are revised through the image quality level revision algorithm to form a pseudo-annotated dataset. Finally, based on the training of the pseudo-labeled data set, the final output result is formed. The experimental results show that the method presented in this paper has good performance and can evaluate most UW-OCTA image quality levels correctly.

References

1. Chollet, F.: Xception: deep learning with depthwise separable convolutions. In: Proceedings of the IEEE Conference on Computer Vision and Pattern Recognition, pp. 1251–1258 (2017)
2. Contributors, M.: MMCV: OpenMMLab computer vision foundation. https://github.com/open-mmlab/mmcv (2018)
3. Dai, L., et al.: A deep learning system for detecting diabetic retinopathy across the disease spectrum. Nat. Commun. **12**(1), 1–11 (2021)
4. Liu, R., et al.: Deepdrid: diabetic retinopathy-grading and image quality estimation challenge. Patterns **3**(6), 100512 (2022)
5. Liu, Z., Mao, H., Wu, C.Y., Feichtenhofer, C., Darrell, T., Xie, S.: A convnet for the 2020s. In: Proceedings of the IEEE/CVF Conference on Computer Vision and Pattern Recognition, pp. 11976–11986 (2022)
6. Madzin, H., Zainuddin, R.: Feature extraction and image matching of 3D lung cancer cell image. In: 2009 International Conference of Soft Computing and Pattern Recognition, pp. 511–515. IEEE (2009)
7. Madzin, H., Zainuddin, R., Mohamed, N.S.: Analysis of visual features in local descriptor for multi-modality medical image. Int. Arab J. Inf. Technol. (IAJIT) **11**(5) (2014)
8. Sheng, B., et al.: An overview of artificial intelligence in diabetic retinopathy and other ocular diseases. Front. Public Health **10**, 971943 (2022)
9. Sheng, B., et al.: Diabetic retinopathy analysis challenge 2022, March 2022. https://doi.org/10.5281/zenodo.6362349
10. Srivastava, N., Hinton, G., Krizhevsky, A., Sutskever, I., Salakhutdinov, R.: Dropout: a simple way to prevent neural networks from overfitting. J. Mach. Learn. Res. **15**(1), 1929–1958 (2014)
11. Tan, M., Le, Q.: Efficientnetv2: smaller models and faster training. In: International Conference on Machine Learning, pp. 10096–10106. PMLR (2021)

12. Tan, Z., Hu, Y., Luo, D., Hu, M., Liu, K.: The clothing image classification algorithm based on the improved xception model. Int. J. Comput. Sci. Eng. **23**(3), 214–223 (2020). https://doi.org/10.1504/ijcse.2020.111426
13. Wang, Z., Bovik, A.C., Sheikh, H.R., Simoncelli, E.P.: Image quality assessment: from error visibility to structural similarity. IEEE Trans. Image Process. **13**(4), 600–612 (2004)
14. Yun, S., Han, D., Oh, S.J., Chun, S., Choe, J., Yoo, Y.: Cutmix: regularization strategy to train strong classifiers with localizable features. In: Proceedings of the IEEE/CVF International Conference on Computer Vision, pp. 6023–6032 (2019)
15. Zhang, H., Cisse, M., Dauphin, Y.N., Lopez-Paz, D.: mixup: beyond empirical risk minimization. arXiv preprint arXiv:1710.09412 (2017)

An Improved U-Net for Diabetic Retinopathy Segmentation

Xin Chen, Yanbin Chen, Chaonan Lin, and Lin Pan[✉]

College of Physics and Information Engineering,
Fuzhou University, Fuzhou 350108, China
panlin@fzu.edu.cn

Abstract. Diabetic retinopathy (DR) is a common diabetic complication that can lead to blindness in severe cases. Ultra-wide (swept
source) optical coherence tomography angiography(UW-OCTA) imaging
can help ophthalmologists in the diagnosis of DR. Automatic and accurate segmentation of the lesion area is essential in the diagnosis of DR.
However, there still remain several challenges for accurately segmenting
lesion areas from UW-OCTA: the various lesion locations, diverse morphology and blurred contrast. To solve these problems, in this paper,
we propose a novel framework to segment neovascularization(NV), nonperfusion areas(NA) and intraretinal microvascular abnormalities(IMA),
which consists of two parts: 1) We respectively input the images of three
lesions into three different channels to achieve three different lesions segmentation at the same time; 2) We improve the traditional 2D U-Net by
adding the residual module and dilated convolution. We evaluate the proposed method on the Diabetic Retinopathy Analysis Challenge (DRAC)
in MICCAI2022. The mean Dice and mean IoU obtained by the method
in the test cases are 0.4757 and 0.3538, respectively.

Keywords: Diabetic Retinopathy · UW-OCTA · Segmentation
Network

1 Introduction

Diabetes is prone to other complications, and diabetic retinopathy (DR) is one of
the most common complications [1]. DR is a chronic progressive disease, which is
caused by microvascular damage in the retina caused by diabetes. DR may lead
to a series of pathological changes such as neovascularization, microangiomas,
and hard exudates, which can cause damage to vision and may lead to blindness
in severe cases. Studies demonstrate that regular DR screening can reduce the
incidence of blindness associated with DR by 90% [2].

Liu [14] et al. found that the quality of fundus images from different devices
varies and that low-quality fundus images increase the training difficulty. Firstly,
normal fundus structures such as blood vessels and optic discs still exist in the
image, which will cause interference to the segmentation. Secondly, the picture

B. Sheng and M. Aubreville (Eds.): MIDOG 2022/DRAC 2022, LNCS 13597, pp. 127–134, 2023.
https://doi.org/10.1007/978-3-031-33658-4_12

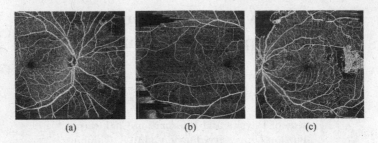

(a)   (b)   (c)

Fig. 1. (a) The fundus vessels occupy most of the image area (b) The image boundary is not clear (c) the terminal morphology of the vessels is different.

quality is uneven. Finally, the diversity of morphology and blurred boundaries also make it hard to accurately segment lesion areas, as shown in Fig. 1.

In recent years, deep learning approaches have gained great popularity and been applied in different fields. In the field of ophthalmology, deep learning has been applied to DR, glaucoma, age-related macular degeneration, and cataracts, four major fundus diseases that can lead to blindness [15]. Ling [3] et al. proposed the AlexNet framework as the basis for microaneurysm detection. Maryam [4] et al. proposed an end-to-end fully convolutional neural network architecture to achieve simultaneous multiple retinal lesions containing exudates, hemorrhages and cotton wool spots. Dai [13] et al. developed a DeepDR system introducing image quality subnetwork and lesion aware subnetwork to improve the diagnostic performance. Good results were achieved in lesion detection of microaneurysms, cotton wool spots, hard exudates and hemorrhages. Kou [5] et al. combined residual modules with recursive convolution based on U-Net for the segmentation of microaneurysms and achieved better results in comparison with other methods such as FCNN and U-Net. Zhang [6] et al. proposed a multi-scale parallel branching network (MPB-CNN) for choroidal neovascularization (CNV) segmentation. Mehdi [7] et al. used traditional morphological methods to segment hard exudates. Zheng [8] et al. developed a new texture segmentation network that first detects candidate capillary nonperfusion regions using an unsupervised segmentation network and then fused texture information for supervised segmentation. This network framework combined the advantages of unsupervised and supervised segmentation techniques and can quantify capillary nonperfusion regions in retinal lesions from different etiologies.

In this work, we provide a fully automated segmentation network. First, the three masks are merged, then the original image and the mask are cropped and chunked using a block-cutting strategy, and then input to the improved U-Net to accurately segment IMA, NA, and NV. In general, the contributions of our work can be summarized in three aspects as follows:

1. We propose an automatic segmentation network evaluated on the Ultra-wide (swept source) optical coherence tomography angiography (UW-OCTA) dataset at the Diabetic Retinopathy Analysis Challenge MICCAI2022.

2. We respectively input the images of three lesions into three different channels to achieve three different lesions segmentation at the same time.
3. In order to overcome the imbalance problem between inter-class, we use the block-cutting strategy to preprocess the data.
4. We improve the traditional 2D U-Net by adding the residual module and dilated convolution(ResHDC U-Net).

2   Methods

To improve the quality of automatic segmentation of different lesions, we propose a modified U-Net network based on the network framework of 2D U-Net [10], in which we change part of the convolution module to null convolution as a way to increase the perceptual field and improve the accuracy of segmentation of lesion regions. Figure 2 shows the overall network framework.

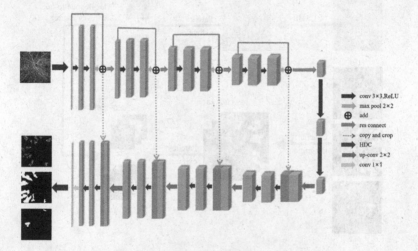

Fig. 2. The automatic segmentation framework.

2.1   Preprocessing

Inspired by [9], in order to improve the contrast of images, we use the Contrast Limited Adaptive Histogram Equalization (CLAHE) to enhance the images. In the dataset, one image may correspond to one or more lesions. Therefore, we integrate the different lesion masks of the same image into channels, and one channel corresponds to one lesion mask. If UW-OCTA image has only mask maps of non-perfused areas but no mask maps of the other two lesions, the values of the corresponding channels are all set to zero. A UW-OCTA image corresponds to a mask block of $1024 \times 1024 \times 3$ in this method.

Since the training result of the neural network will depend on the number of training data, we cut the images in the training data to increase the training samples and avoid the overfitting phenomenon. From the masks of MA and NV, it can be concluded that the foreground region is much smaller than the background region, which easily leads to the imbalance between inter-class. Therefore, we propose a sample method to crop the 1024×1024 image blocks into 256×256 image blocks at a step size of 0.5. When the number of foreground pixel values in the mask block is greater than 0, it corresponds to the original image, otherwise it is discarded. This is helpful to reduce the ratio between the area of the focal region and the non-focal region to overcome the interclass imbalance problem. After that, a series of enhancement processing such as random rotation and mirror flip are performed to increase the number of data samples. Figure 3 shows the whole preprocessing process.

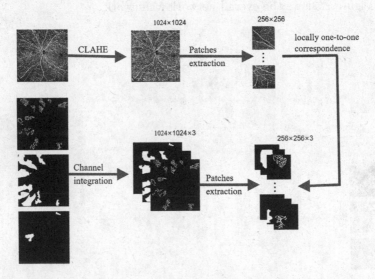

Fig. 3. The preprocessing process.

2.2   ResHDC U-Net

U-Net [10] is proposed and applied to cell segmentation in 2015 and has been widely used in different classes of medical image segmentation since then. The main feature of U-Net was the incorporation of jump connections between the encoder and decoder, and it is named because of the 'U'-like shape of its architecture. U-Net uses upsampling and splicing jump connections to synthesize multi-scale feature information, but further improvements are needed in UW-OCTA image segmentation.

Residual Structure. As the layers of the neural networks are deepened, the risk of gradient disappearance or gradient explosion also increases. Therefore, we add the residual structure [11] to the network, as shown in Fig. 4.

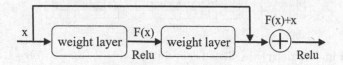

Fig. 4. Residual structure.

In Fig. 4, x represents the input and $F(x)$ represents the output of the residual block before passing through the activation function of the next layer, as shown in Eq. 1.

$$F(x) = W_2\delta(W_1x) \tag{1}$$

where W_1 and W_2 represent the weights of the first and second layers, respectively, and δ represents the activation function. Then the output of the final residual module is $\delta(F(x) + x)$.

We impose the residual module in front of each pooling layer of the encoder structure to ensure that shallow information can be passed to deep layers and deep errors can be fed back to shallow layers as the network layers of the encoder structure deepen. This strategy allows the input to be connected to the output of each convolutional layer in the encoder, solving the problem of gradient disappearance and gradient explosion.

Dilated Convolution. The difference from normal convolution is that the dilated convolution adds holes to the convolution kernel, which simply means inserting zeros between the pixel values of the normal convolution kernel to increase the expansion rate r of the network, as shown in Fig. 5. Dilated convolution can increase the perceptual field to capture multi-scale contextual information without decreasing the image resolution and losing image information. The size of the receptive field can be adjusted by the expansion rate r. For a 2D image, the dilated convolution process for pixel i on the output feature layer y is as follows:

$$y(i) = \sum_k x(i + rk)w(k) \tag{2}$$

w is the convolution kernel, k is the convolution kernel size, x is the convolution input, and r is the expansion rate.

Due to the convolution of voids of the same size r is superimposed several times, the continuity of information may be lost and leads to the possibility of irrelevant information at a distance. Therefore, we use a hybrid dilated convolution(HDC) [12] algorithm, replacing some of the convolution and pooling layers in the U-Net network with serial dilated convolutions of expansion rates of 1, 3, and 5, respectively. This can retain more image detail information and improve the generalization ability of the network.

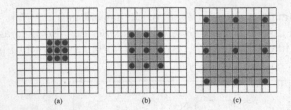

Fig. 5. Principle of dilated convolution(a)$r=1$, (b)$r=2$, (c)$r=4$.

3   Experimental Results

3.1   Dataset

Ultra-wide OCTA can detect changes in DR neovascularization non-invasively, and is therefore an important imaging modality to help ophthalmologists diagnose PDR. Therefore, we used the UW-OCTA dataset provided by DRAC MICCAI2022 for testing the effectiveness of the algorithm.

3.2   Implementation Details

As an experimental setting, we choose PyTorch to implement our model and use an NVIDIA Tesla P100 16 GB GPU for training. The input size of the network is 256×256 and the batch size is 4. In our model, we set the epoch to 200, the initial learning rate is 1×10^{-2}, and use dice loss as the loss function.

3.3   Evaluation Metrics

We use the Dice and IoU provided by DRAC 2022 as the main evaluation criteria to evaluate the segmentation performance.

$$Dice = \frac{2|X \cap Y|}{|X| + |Y|} \tag{3}$$

$$IoU = \frac{X \cap Y}{X \cup Y} \tag{4}$$

X and Y denote the two sets of true and predicted values, respectively. $|X|$ and $|Y|$ represent the number of X and Y elements.

3.4   Results

To evaluate the effectiveness of the proposed method, we compare U-Net with our method. At the same time, the visual comparison is carried out in the case of the same dataset and data parameters. In addition, to explore the advantages of our method, we validate our method on the UW-OCTA test set using the evaluation metrics Dice and IoU.

It can be seen from Table 1 that compared with U-Net, our proposed method has better performance on Dice and IoU. The Dice and IoU of our method on the test set are 0.4757 and 0.3538, respectively (Fig. 6).

Table 1. Dice score of the proposed method and other baseline methods on the validation set.

Method	IoU			Dice		
	IMA	NA	NV	IMA	NA	NV
U-Net	0.1664	0.4309	0.2574	0.2686	0.5729	0.3669
Ours	0.2220	0.5080	0.3314	0.3289	0.6470	0.4511

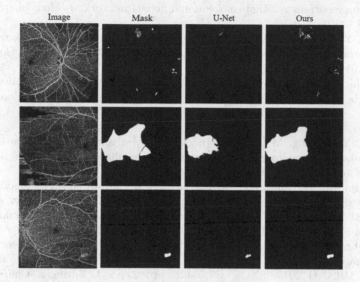

Fig. 6. Visualization results of different methods on the validation set.

4 Conclusion

In this paper, we propose a method for fully automatic segmentation of three types of fundus lesions simultaneously. Firstly, the masks of the three lesions are integrated and segmented, and then the ResHDC U-Net network is used to achieve the final segmentation. The proposed method has been evaluated on the dataset of DRAC MICCAI2022. The experimental results show that the method is helpful for diabetic retinopathy segmentation.

References

1. Andersen, J.K.H., Hubel, M.S., Rasmussen, M.L., et al.: Automatic detection of abnormalities and grading of diabetic retinopathy in 6-field retinal images: integration of segmentation into classification. Transl. Vis. Sci. Technol. **11**(6), 19–19 (2022)
2. Busbee, B.: The 25-year incidence of visual impairment in type 1 diabetes mellitus: the wisconsin epidemiologic study of diabetic retinopathy. Evid.-Based Ophthalmol. **12**(1), 28–29 (2011)

3. Dai, L., Fang, R., Li, H., et al.: Clinical report guided retinal microaneurysm detection with multi-sieving deep learning. IEEE Trans. Med. Imaging **37**(5), 1149–1161 (2018)
4. Badar, M., Shahzad, M., Fraz, M.M.: Simultaneous segmentation of multiple retinal pathologies using fully convolutional deep neural network. In: Nixon, M., Mahmoodi, S., Zwiggelaar, R. (eds.) MIUA 2018. CCIS, vol. 894, pp. 313–324. Springer, Cham (2018). https://doi.org/10.1007/978-3-319-95921-4_29
5. Kou, C., Li, W., Liang, W., et al.: Microaneurysms segmentation with a U-Net based on recurrent residual convolutional neural network. J. Med. Imaging **6**(2), 025008 (2019)
6. Zhang, Y., Ji, Z., Wang, Y., et al.: MPB-CNN: a multi-scale parallel branch CNN for choroidal neovascularization segmentation in SD-OCT images. OSA Continuum **2**(3), 1011–1027 (2019)
7. Eadgahi, M.G.F., Pourreza, H.: Localization of hard exudates in retinal fundus image by mathematical morphology operations. In: 2012 2nd International eConference on Computer and Knowledge Engineering (ICCKE), pp. 185–189. IEEE (2012)
8. Zheng, Y., Kwong, M.T., MacCormick, I.J.C., et al.: A comprehensive texture segmentation framework for segmentation of capillary non-perfusion regions in fundus fluorescein angiograms[J]. PloS one **9**(4), e93624 (2014)
9. Zulfahmi, R., Noviyanti, D.S., Utami, G.R., et al.: Improved image quality retinal fundus with contrast limited adaptive histogram equalization and filter variation. In: 2019 International Conference on Informatics, Multimedia, Cyber and Information System (ICIMCIS), pp. 49–54. IEEE (2019)
10. Ronneberger, O., Fischer, P., Brox, T.: U-net: convolutional networks for biomedical image segmentation. In: Navab, N., Hornegger, J., Wells, W.M., Frangi, A.F. (eds.) MICCAI 2015. LNCS, vol. 9351, pp. 234–241. Springer, Cham (2015). https://doi.org/10.1007/978-3-319-24574-4_28
11. He, K., Zhang, X., Ren, S., et al.: Deep residual learning for image recognition. In: Proceedings of the IEEE Conference on Computer Vision and Pattern Recognition, pp. 770–778 (2016)
12. Wang, P., Chen, P., Yuan, Y., et al.: Understanding convolution for semantic segmentation. In: 2018 IEEE Winter Conference on Applications of Computer Vision (WACV), pp. 1451–1460. IEEE (2018)
13. Dai, L., Wu, L., Li, H., et al.: A deep learning system for detecting diabetic retinopathy across the disease spectrum. Nat. Commun. **12**(1), 1–11 (2021)
14. Liu, R., Wang, X., Wu, Q., et al.: DeepDRiD: diabetic retinopathy-grading and image quality estimation challenge. Patterns **3**(6), 100512 (2022)
15. Sheng, B., Chen, X., Li, T., Ma, T., Yang, Y., Bi, L., Zhang, X.: An overview of artificial intelligence in diabetic retinopathy and other ocular diseases. Front. Public Health **10**, 971943 (2022). https://doi.org/10.3389/fpubh.2022.971943

A Vision Transformer Based Deep Learning Architecture for Automatic Diagnosis of Diabetic Retinopathy in Optical Coherence Tomography Angiography

Sungjin Choi[1]([⊠])[iD], Bosoung Jeoun[2][iD], Jaeyoung Anh[1][iD], Jaehyup Jeong[2][iD], Yongjin Choi[1][iD], Dowan Kwon[1][iD], Unho Kim[1][iD], and Seoyoung Shin[1][iD]

[1] AI/DX Convergence Business Group, KT, Seongnam-si, Korea
{sj717.choi,ahn.tea,yongjin7.choi,kwon.dowan,unho.kim,
shin.seoyoung}@kt.com
[2] AI2XL Research Center, KT, Seongnam-si, Korea
{lynn.jeoun,jaehyup.jeong}@kt.com

Abstract. The diabetic retinopathy (DR) is an eye abnormality that highly causes blindness or affects majority of patients with a history of 15 years of diabetes at least. To diagnose DR, image quality assessment, lesion segmentation, and DR grade classification are required. However, any automatic DR analysis has not been developed yet. Therefore, the challenge DRAC 2022 suggested three tasks; Task 1: Segmentation of Diabetic Retinopathy Lesions; Task 2: Image Quality Assessment; Task 3: Diabetic Retinopathy Grading. These tasks aim to be built robust but adaptable model for automatic DR diagnosis with provided OCT angiography (OCTA) dataset. In this paper, we proposed an automatic DR diagnosis method with deep learning benchmarking, and image processing from OCTA. The proposed method achieved Dice of 0.6046, Cohen kappa of 0.8075, and 0.8902 for each task respectively with the second place ranking in the competition. The code is available at https://github.com/KT-biohealth/DRAC22.

1 Introduction

Diabetic retinopathy (DR) is Eye disease frequently observed in diabetic patients. In serious cases, it is a major cause of vision loss [11–13]. DR is diagnosed with visual examination of retinal fundus images to identify one or more lesions such as microaneurysm and hemorrhages present. DR grade can be classified into non-proliferative DR (NPDR) and proliferative DR (PDR) [12]. NPDR is an early stage of DR, which is possible to check by microaneurysms. On the other hand, PDR is an advanced stage of DR and periodic inspections are required. To detect DR lesions, there are several methods such as OCT angiography (OCTA) which has the visualization function in microvascular level,

B. Sheng and M. Aubreville (Eds.): MIDOG 2022/DRAC 2022, LNCS 13597, pp. 135–145, 2023.
https://doi.org/10.1007/978-3-031-33658-4_13

and SS(swept-source)-OCTA, UW(ultra-wide)-OCTA. Among techniques, UW-OCTA is a useful technique that can detect changes in DR neovascularization without invasiveness [14]. However, since UW-OCTA can cause problems such as artifact and clarity, it is necessary to verify image quality. After verifying images, lesion segmentation and DR Grade analysis must be performed, so a simple screening solution that can quantitatively make a subjective diagnosis is needed [13]. In the meantime, various studies on DR grade have been conducted, and the screening solution must automatically determine the OCT image quality and include segmentation and DR Grade measurement functions for diagnosable images. In this paper, an automatic diagnosis of OCTA through benchmarking various deep learning networks, image processing, and algorithms for each task was developed.

2   Data and Preprocessing

We have used the DRAC 2022 dataset which was provided from MICCAI Challenge 2022. The dataset consisted of three different classes of OCTA images for suggested task. Since OCTA images are grayscale and contain lots of noisy speckle patterns, the image enhancement to improve the performance of deep learning [1]. Thus, the image filtering, contrast limited adaptive histogram equalization (CLAHE), and colorization were applied to improve image quality. A bilateral filter was used in order to suppress the noisy patterns of the input image. As shown in Fig. 1b, the smoothened image was processed by the OpenCV bilateral filter with 50 sigma color and 5 sigma space [2]. The optimized parameters of bilateral filter were chosen that could simultaneously reduce noise and maintain important structures such as blood vessels through experiments. The blood vessels were observed brighter than other structures in OCTA images. To emphasize the features of blood vessels, we developed a coloring method that is explained in the following theorem Fig. 1(c).

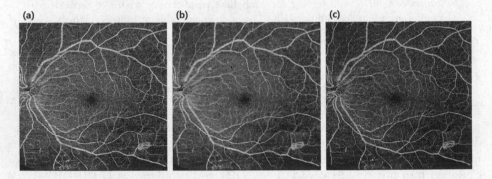

Fig. 1. Pre-processing result of 097.png for both training and inference (a) an original image (b) an enhanced image (c) a colored image.

Theorem 1. *Colorization*

OutputImage [0] = inputImage

OutputImage [1] = inputImage (pixel only greater than Threshold)

OutputImage [2] = inputImage (pixel only smaller than Threshold)

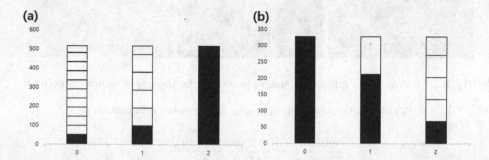

Fig. 2. The numbers of classes for classification (a) task 2 (b) task 3; the black boxes mean the original numbers of each class and the white boxes mean the increased numbers of images.

In addition, it tends to be difficult to gather illness or problem data in a medical environment. The given data set of Task 2 and Task 3 for training was observed to be also imbalanced, as is shown in Fig. 2. Since the classification performance under an imbalanced dataset is known as low, we tried to make every class have the same number of data using the upper sampling method. Each image of the fewer class was randomly processed and copied repeatedly until it became the largest number. We used vertical flip, horizontal flip, rotation, zooming, sharpening, and gamma correction as the image processing methods. Figure 3 shows the examples generated by the upper sampling. The upper sampling method was used before the training, and the generated images were saved as an image format file.

3 Models and Training

3.1 Segmentation

The several experiments were implemented to seek optimal deep learning network for segmentation task. Among those, three networks were chosen which have shown the best results; the convolution based network, ConvNext [7], the transformer based network, SegFormer [8], and Swin Transformer [9]. We ensembled the best performed networks to utilize the benefits of both convolution and transformer based network. The pre-trained models were used for each network with the best hyperparameter combinations of various criterions (Table 1). The original input size, 1024×1024, was used to prevent feature loss. The AdamW and polyLR were selected for optimizer and scheduler respectively. The vertical and horizontal flip, random rotation and CLAHE [10] were for augmentation.

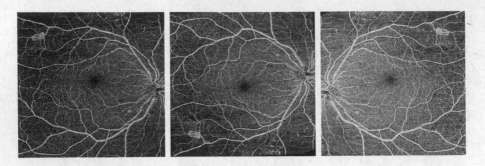

Fig. 3. The example of additional images generated by the upper sampling method.

Table 1. Segmentation networks and training information.

Network	Criterion	Optimizer
ConvNext	CE, Dice	AdamW
SegFormer	Focal, Dice	AdamW
Swin	Focal, Dice	AdamW

Class1 (Intraretinal Microvascular Abnormalities) and Class3 (Neovascularization). It was observed that the data dispersion of both class 1 and class 3 were relatively rare compare to class 2. Thus, in the case of managing the overlapped areas between class 1 and 2, the priority was given to class 1 than class 2 during training. Likewise, class 3 had priority over class 2 for overlapped areas between class 2 and 3. To achieve the best performance, the original image from Task 3 and the image of class 1 and 3 from Task 1 were fused by adding the image features. Furthermore, the pseudo labeling method was implemented. The best performed network of Task 1 inferenced the original image from Task 3 to generate the masks of each class. The results of pseudo labeling methods were used as input again for training. The final segmentation results were obtained through post processing the results of ensembled networks morphologically with logical OR calculation (Fig. 4).

Class 2 (Nonperfusion Areas). Due to class 1 and 3, the area of class 2 seemed partly void. To redeem the missing area, class 2 became the prior class than class 1 and 3 to manage overlapped area with class 1 and 3. The processed mask data for class 2 and the mask with class 2 only were also used for training while using normal masks which have priority to class 1 and 3. The final segmentation results were obtained with same method as class 1 and 3 but using logical AND instead of logical OR calculation (Fig. 5).

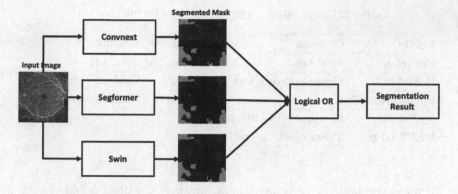

Fig. 4. The overview of segmentation strategy for class 1 and class 3.

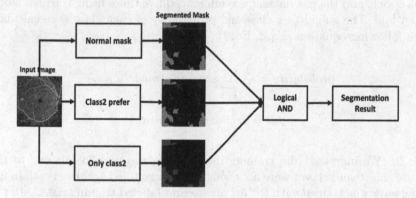

Fig. 5. The overview of segmentation strategy for class 2.

3.2 Classification

We selected several image classification networks based on the benchmark (Table 2). Although we considered higher ranked networks at first, we only utilized some networks whose pre-trained model files could be found easily. We experimented with various criterions, optimizations, and scheduler combinations. The combination marked as bold in Table 2 was our final selection; Beit, cross entropy loss, ADAMW, and Step LR [3]. The argument of training images included vertical and horizontal flip, and both training and testing images were normalized by the maximum intensity of the OCTA image.

Table 2. Classification networks and training information.

Network	Criterion	Optimizer	Scheduler
ResNet 101D	MSE loss	Adam	**Step LR**
DenseNet	**Cross entropy loss**	RAdam	Gradual warm up
EfficientNet B7	Focal loss	**AdamW**	Cosine annealing LR
NFNet F6	Bi-tempered logistic loss	SAM	–
BEIT large	Taylor cross entropy loss	–	–

To make the best use of training data, we used stratified 5-fold cross validation. The final model of each fold in the training session was the best Cohen kappa epoch, and the test dataset was inferenced five times using the final models of each fold. The submitting class and probability of each class were calculated as the following equation (Eq. 1, Eq. 2).

$$probability = \frac{1}{N} \sum_{i=1}^{N} softmax(model_{output}) \tag{1}$$

$$label = argmax(probability) \tag{2}$$

Task 2. We increased the training image using the pseudo labeling method because class 0 and class 1 were not enough to be identified accurately [4]. In first, the network was trained with the imbalance and labeled training data. After the trained network had proper classification performance, a pseudo-labeled dataset was created by the inference of the testing dataset. Second training was performed including the image and label of class 0 and class 1 among the generated pseudo-label dataset (Fig. 6).

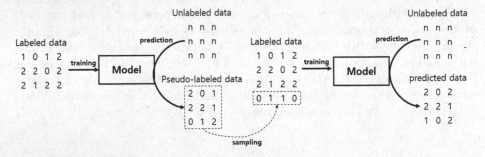

Fig. 6. Classification strategy for task 2: pseudo labeling.

Although there would be error labels in the created pseudo-label, the second training improved the classification performance compared to the result of the first training. To achieve better results, an ensemble between BEIT and NFNet was also used, as in the following equation (Eq. 3, Eq. 4).

$$probability = 0.55 \times softmax(model_{output_{beit}}) + 0.45 \times softmax(model_{output_{nfnet}}),$$
$$\tag{3}$$
$$label = argmax(probability) \tag{4}$$

Task 3. In task 3, the grade of DR was labeled based on the presence of retinal lesions such as intraretinal microvascular abnormalities (IRMAs), nonperfusion areas and neovascularization for image classification, which was identified as class 0, 1, and 2 [5]. The neovascularization, class 2, is especially known as the lesion for diagnostic proliferative diabetic retinopathy [6]. We designed a combination of classification and segmentation to grade diabetic retinopathy with OCTA images and mask images of these retinal lesions (Fig. 7).

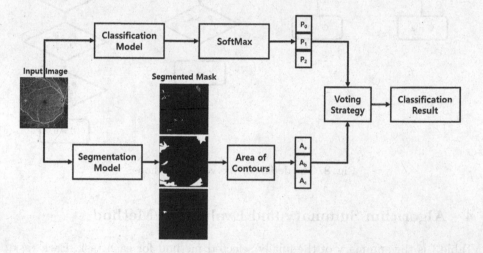

Fig. 7. Classification strategy for task 3: voting with segmentation.

The segmentation model was trained using training images and lesion masks from task 1, and then the model made three binary lesion images for each input image. Aa, Ab, and Ac mean automatically measured areas of IRMAs, nonperfusion, and neovascularization respectively. P0, P1, and P2 mean the probability of class 0, 1, and 2 respectively, and the probability values are the SoftMax output that is generated by the trained classification model. Based on the rule illustrated in Fig. 8, the test image was finally classified using the probabilities of each class and the areas of each lesion. If the area of neovascularization was large and the probability of class 2 was high, the classification result would be determined as class 2. In addition, if the original result of classification was class 0 but the area of IRMAs was large, the result would be changed to class 1. If any lesion masks did not exist in the segmentation result, the classification result was changed to class 0.

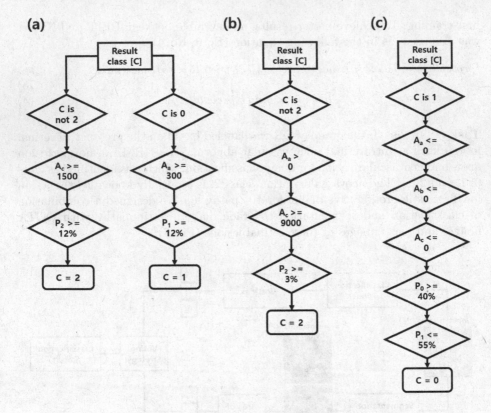

Fig. 8. The details of the voting methods.

4   Algorithm Summary and Evaluation Method

Table 3 is the summary of the finally selected method for each task. Each task was evaluated with presented evaluation metrics from MICCAI DRAC 2022 Challenge. The mean of Dice Similarity Coefficient (mDice) and Intersection of Union (mIoU) were calculated for Task 1, and quadratic weighted kappa (Cohen kappa) and Area Under Curve (AUC) were used for both Task 2, and 3 for the competition.

5   Results

5.1   Task 1

The results of Task 1 are shown Fig. 9(a). The mDice result of ConvNext without coloring was 0.4789 then it was increased to 0.5021 after using colorization data. The result of fusing classes and using pseudo labeling method was 0.5352 and 0.5884, respectively. As the result of including the network ensemble, the final mDice achieved 0.6046.

Table 3. The summary of each task.

	Task1	Task2	Task3
Upper sampling	No	No	Yes
Preprocessing	Coloring	Bilateral Filtering	Coloring
Augmentation	V Flip, H Flip, Rotation, CLAHE	V Flip, H Flip	V Flip, H Flip
Normalization	Mean and Std of Training set	Divided by 255	Divided by 255
Model architecture	SegFormer, ConvNeXt, Swin	Beit, NFNet	Beit
Postprocessing	Pseudo Labeling, Model Ensemble	Pseudo Labeling, Model Ensemble	Voting with segmentation

5.2 Task 2

When we trained with the original unbalanced data set, Cohen kappa was 0.7410. The Cohen kappa was increased to 0.7474 with the image coloring method for both training and testing datasets. Those previous two Cohen kappa values are the result of a 5-fold ensemble. We used the whole training data without the ensemble, and then the classification performance increased to 0.7674. The Cohen kappa result with pseudo-labeling was 0.8016. Our final Cohen kappa score was 0.8075, which was the result from the ensemble of BEIT and NFNet (Fig. 9(b)).

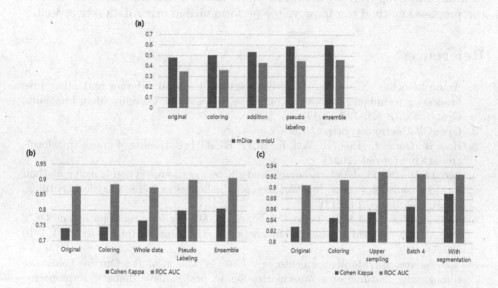

Fig. 9. The Result values of each method (a) task 1 (b) task 2 (c) task 3.

5.3   Task 3

When we trained with the original unbalanced data set, Cohen kappa was 0. 8279. Image pre-processing helped to improve the performance of classification. Image preprocessing helped to improve the performance of classification. In detail, the Cohen kappa score was 0.8444 with the image coloring method and 0.8558 with the upper sampling method. Through hyper parameter optimization, we found that batch number 4 is better than 2. The Cohen kappa of batch 4 was 0.8658. Finally, we achieved 0.8902 of the Cohen kappa using the combining method of classification and segmentation (Fig. 9(c)).

6   Conclusion

In this paper, we have proposed a method to improve the performance of a given data set using existing networks. It avoids complex method, leaded to good performance with simple method. To overcome the small amount of training data, upper sampling, pseudo-labeling, and augmentation were used for each task. Also, performance improvement was achieved by ensemble or voting using multiple networks. The proposed method has improved performance compared to using only deep learning networks. In the future work, we will check whether the proposed method can improve the performance in other data sets as well.

References

1. Anantrasirichai, N., et al.: Adaptive-weighted bilateral filtering and other pre-processing techniques for optical coherence tomography. Comput. Med. Imagining Graph. **38**(6), 526–539 (2014)
2. OpenCV Document. https://docs.opencv.org
3. Bao, H., Dong, L., Piao, S., Wei, F.: BEiT: BERT pre-training of image transformers. arXiv preprint (2021)
4. Lee, D.-H.: Pseudo-label: the simple and efficient semi-supervised learning method for deep neural networks. In: Workshop on Challenges in Representation Learning, ICML, vol. 3. no. 2 (2013)
5. de Barros Garcia, J.M.B., Isaac, D.L.C., Avila, M.: Diabetic retinopathy and OCT angiography: clinical findings and future perspectives. Int. J. Retin. Vitr. **3**(1), 1–10 (2017)
6. Vaz-Pereira, S., Morais-Sarmento, T., Esteves Marques, R.: Optical coherence tomography features of neovascularization in proliferative diabetic retinopathy: a systematic review. Int. J. Retin. Vitr. **6** (2020). Article number: 26. https://doi.org/10.1186/s40942-020-00230-3
7. Liu, Z., Mao, H., Wu, C.-Y., Feichtenhofer, C., Darrell, T., Xie, S.: A convent for the 2020s. arXiv (2022). arXiv:2201.03545
8. Xie, E., Wang, W., Yu, Z., Anandkumar, A., Alvarez, M.J., Luo, P.: SegFormer: simple and efficient design for semantic segmentation with transformers. arXiv (2021). arXiv:2105.15203
9. Liu, Z., et al.: Swin transformer: hierarchical vision transformer using shifted windows. arXiv (2021). arXiv:2103.14030

10. Zuiderveld, K.: Contrast limited adaptive histogram equalization. In: Graphics Gems IV, pp. 474–485 (1994)
11. Dai, L., Wu, L., Li, H., et al.: A deep learning system for detecting diabetic retinopathy across the disease spectrum. Nat. Commun. **12**(1), 1–11 (2021)
12. Liu, R., Wang, X., Wu, Q., et al.: DeepDRiD: diabetic retinopathy-grading and image quality estimation challenge. Patterns **3**(6), 100512 (2022)
13. Sheng, B., et al.: An overview of artificial intelligence in diabetic retinopathy and other ocular diseases. Front. Public Health **10**, 971943 (2022). https://doi.org/10.3389/fpubh.2022.971943
14. https://drac22.grand-challenge.org/

Segmentation, Classification, and Quality Assessment of UW-OCTA Images for the Diagnosis of Diabetic Retinopathy

Yihao Li[1,2(✉)], Rachid Zeghlache[1,2], Ikram Brahim[1,2], Hui Xu[1], Yubo Tan[3],
Pierre-Henri Conze[1,4], Mathieu Lamard[1,2], Gwenolé Quellec[1],
and Mostafa El Habib Daho[1,2(✉)]

[1] LaTIM UMR 1101, Inserm, Brest, France
`Yihao.Li@etudiant.univ-brest.fr`, `mostafa.elhabibdaho@univ-brest.fr`
[2] University of Western Brittany, Brest, France
[3] University of Electronic Science and Technology of China, Chengdu, China
[4] IMT Atlantique, Brest, France

Abstract. Diabetic Retinopathy (DR) is a severe complication of diabetes that can cause blindness. Although effective treatments exist (notably laser) to slow the progression of the disease and prevent blindness, the best treatment remains prevention through regular checkups (at least once a year) with an ophthalmologist. Optical Coherence Tomography Angiography (OCTA) allows for the visualization of the retinal vascularization, and the choroid at the microvascular level in great detail. This allows doctors to diagnose DR with more precision. In recent years, algorithms for DR diagnosis have emerged along with the development of deep learning and the improvement of computer hardware. However, these usually focus on retina photography. There are no current methods that can automatically analyze DR using Ultra-Wide OCTA (UW-OCTA). The Diabetic Retinopathy Analysis Challenge 2022 (DRAC22) provides a standardized UW-OCTA dataset to train and test the effectiveness of various algorithms on three tasks: lesions segmentation, quality assessment, and DR grading. In this paper, we will present our solutions for the three tasks of the DRAC22 challenge. The obtained results are promising and have allowed us to position ourselves in the TOP 5 of the segmentation task, the TOP 4 of the quality assessment task, and the TOP 3 of the DR grading task. The code is available at https://github.com/Mostafa-EHD/Diabetic_Retinopathy_OCTA.

Keywords: Diabetic Retinopathy Analysis Challenge · UW-OCTA · Deep Learning · Segmentation · Quality Assessment · DR Grading

This work is supported by the ANR RHU project Evired. This work benefits from State aid managed by the French National Research Agency under "Investissement d'Avenir" program bearing the reference ANR-18-RHUS-0008.

B. Sheng and M. Aubreville (Eds.): MIDOG 2022/DRAC 2022, LNCS 13597, pp. 146–160, 2023.
https://doi.org/10.1007/978-3-031-33658-4_14

1   Introduction

Diabetes, specifically uncontrolled diabetes causing Diabetic Retinopathy (DR), is among the leading causes of blindness. DR is a condition that affects approximately 78% of people with a history of diabetes of 15 years or more [23]. In the early stages, DR is considered a silent disease. For this reason, seeing an ophthalmologist regularly, especially if you have diabetes, is essential to avoid the risk of serious complications, including blindness.

DR is diagnosed by visually inspecting fundus images for retinal lesions such as Microaneurysms (MA), Intraretinal Microvascular Anomalies (IRMA), areas of Non-Perfusion, and Neovascularization. Fundus photography and fundus fluorescein angiography (FFA) are the two most commonly used tools for DR screening. Traditional diagnosis of DR relies mainly on these two modalities, especially for Proliferative Diabetic Retinopathy (PDR), which seriously endangers visual health. However, fundus photography has difficulty detecting early or small neovascular lesions, and FFA is an invasive fundus imaging that cannot be used in allergic patients, pregnant women, or those with impaired liver and kidney function.

Optical Coherence Tomography Angiography (OCTA) is a new non-invasive imaging technique that generates volumetric angiographic images in seconds. It can display both structural, and blood flow information [3]. Due to the quantity and quality of the information provided by this modality, OCTA is being increasingly used for diagnosing DR at the early stages. In addition, the Swept-Source OCTA (SS-OCTA) allows the individual assessment of choroidal vascularization and the Ultra-Wide Optical Coherence Tomography (UW-OCTA) imaging modality has shown a more significant pathological burden in the retinal periphery that was not captured by typical OCTA [26].

Several DR diagnosis algorithms have emerged in recent years through improved computer hardware, deep learning, and data availability [1,2,8,9,13–15,19,25]. Some works have already used SS-OCTA to assess the qualitative characteristics of DR [18] and others have used UW-OCTA on DR analysis [26] [8,17]. However, there is currently no work that can automatically analyze DR using UW-OCTA. In the DR analysis process, the image quality of the UW-OCTA must first be evaluated, and the best quality images are selected. Then, DR analysis, such as lesion segmentation and PDR detection, is performed. Therefore, it is crucial to build a full pipeline to perform automatic image quality assessment, lesion segmentation, and PDR detection.

The Diabetic Retinopathy Analysis Challenge 2022 (DRAC22) provides a standardized UW-OCTA dataset to train and test the effectiveness of various algorithms.

DRAC22 is a first edition associated with MICCAI 2022 that offers three tasks to choose from:

- Task 1: Segmentation of DR lesions.
- Task 2: Image quality assessment.
- Task 3: Classification of DR.

This article will present our three proposed solutions to solve each task of the DRAC22 challenge.

2   Materials and Methods

2.1   Datasets

The instrument used to gather the dataset in this challenge was an SS-OCTA system (VG200D, SVision Imaging, Ltd., Luoyang, Henan, China), which works near 1050nm and features a combination of industry-leading specifications, including an ultrafast scan speed of 200,000 AScans per second [1][2].

The following table summarizes the data collected by the DRAC22 Challenge organizers. All images are 2D en-face images.

Table 1. DRAC 2022 datasets

Task	# Training images	# Test images
Task 1 - Segmentation	109	65
Task 2 - Quality assessment	665	438
Task 3 - Classification	611	386

The training set consists of 109 images and corresponding labels for the first task. The dataset, as shown in Fig. 1a, contains three different Diabetic Retinopathy Lesions: Intraretinal Microvascular Abnormalities (1), Nonperfusion Areas (2), and Neovascularization (3). The test set consists of 65 images.

For the second task, quality assessment, the organizers propose a dataset of 665 and 438 images for training and testing, respectively. These images (see Fig. 1b) are grouped into three categories: Poor quality level (0), Good quality level (1), and Excellent quality level (2).

The third dataset is dedicated to the classification task. It contains 611 images for learning and 386 for testing, grouped into three different diabetic retinopathy grades as shown in Fig. 1c: Normal (0), NPDR (1), and PDR (2).

---

[1] https://drac22.grand-challenge.org/Data/.

[2] https://svisionimaging.com/index.fluorescein/en_us/home/.

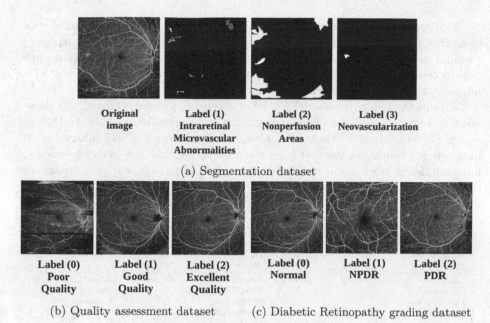

(a) Segmentation dataset

(b) Quality assessment dataset   (c) Diabetic Retinopathy grading dataset

Fig. 1. DRAC22 Challenge Dataset

2.2   Task 1 - Segmentation

In this section, we introduce the models and techniques used to solve the segmentation problem: nnU-Net and V-Net.

U-Net is a simple and successful architecture that has quickly become a standard for medical image segmentation [16]. However, adapting U-Net to new problems is not straightforward, as the exact architecture and the different parameters would have to be chosen. The no-new-U-Net (nnU-Net) method provides an automated end-to-end pipeline that can be trained and inferred on any medical dataset for segmentation [6]. nnU-Net outperformed state-of-the-art architectures in the Medical Decathlon Challenge[3].

After analyzing the dataset of the segmentation task, we noticed that the three labels overlapped on several images, so we opted to train three binary segmentation models for the segmentation of each of the labels.

We first trained an nnU-Net model for each label. Initial results showed that the trained nnU-Net for label 2 (nonperfusion areas) performed well. However, the other two models trained on label 1 (intraretinal microvascular anomalies) and label 3 (neovascularization) did not learn well, and the results were poor.

---

[3] http://medicaldecathlon.com/results/.

As a second solution, we trained a binary V-Net model for the segmentation of each of the labels: 1 and 3. V-Net is a U-Net-based network that incorporates residual blocks into the network. Residual linking encourages the training process to converge faster [12]. The hyperparameters can be continuously tested to improve segmentation performance over nnU-Net.

We observed that the model for label 3 tended to over-segment. To alleviate this issue, we added a classification step to predict the probability that an image contains label 3. The added classifier (based on ResNet) allowed us to improve our results on the test base since an image that has been classified as not containing label 3 will not be segmented. The models and parameters are summarized in Table 2.

Table 2. Implementations of different labels

Segmentation Task	Architecture	Image Size	Data Augmentation	Loss	Optimizer
Label 1 - Intraretinal Microvascular Abnormalities	2D V-Net	1024×1024	RandomAffine Rand2DElastic Default Data Augmentation	Dice loss	Adam lr = 1e-3 ExponentialLR (gamma = 0.99)
Label 2 - Nonperfusion Areas	2D nnU-Net	1024×1024 Normalization	Random rotations, Random scaling, Random elastic Deformations, Gamma correction and mirroring	Dice loss + Cross-entropy loss	Adam lr = 0.01
Label 3 - Neovascularization	2D V-Net	1024×1024	RandomAffine Rand2DElastic Default Data Augmentation	Dice loss	Adam lr = 1e-3 ExponentialLR (gamma = 0.99)
Label 3 Classifier	ResNet101	1024×1024	Default Data Augmentation	Cross-entropy loss	Adam lr = 1e-4 weight_decay=1e-4 ExponentialLR (gamma = 0.99)

Dice and cross-entropy loss are used to train the nnU-Net network. The dice loss formulation is adapted from the variant proposed in [6]. It is implemented as follows:

$$L_{dice} = -\frac{2}{|K|} \sum_{k \in K} \frac{\sum_{i \in I} u_i^k v_i^k}{\sum_{i \in I} u_i^k + \sum_{i \in I} v_i^k} \tag{1}$$

where u is the softmax output of the network and v is the one hot encoding for the ground truth segmentation map. Both u and v have shape $I \times K$ with $i \in I$ being the number of pixels in the training patch/batch and $k \in K$ being the classes.

Adam optimizer is used to train the nnU-Net network with an initial learning rate of 0.01. A five-fold cross-validation procedure is used, and the model has been trained over 1000 epochs per fold with a batch size fixed to 2. Whenever the exponential moving average of the training losses did not improve by at least 5×10^{-3} within the last 30 epochs, the learning rate was reduced by factor 5. The training was stopped automatically if the exponential moving average of the validation losses did not improve by more than 5×10^{-3} within the last 60 epochs, but not before the learning rate was smaller than 10^{-6} [6].

Dice loss (include_background = False) is used to train the V-Net network. Besides the default data augmentation (random crop, random flip, and random rotation), RandomAffine and Rand2DElastic were also used. And mean_Dice is used to select the best checkpoint. The training epoch is 1000, the optimizer is Adam, and the batch size is 3. Finally, for the classifier of label 3, the batch size is 4 and the epoch is 500.

2.3 Task 2 and 3 - Quality Assessment and Classification of DR

As both Task 2 and Task 3 involved three-labels classifications, the pipeline was the same. To verify the performance of the different models, we used five-fold cross-validation to test six architectures (17 backbones): ResNet [4], DenseNet [5], EfficientNet [22], VGG [21], ConvNeXt [11], Swin-Transformer [10].

These backbones were pre-trained using ImageNet and originated from the timm library [24]. We kept the original size of the image during training, i.e., 1024×1024.

We have performed several experiments to improve our models using different optimizers (SGD, Adam, AdamW, RMSprop), schedulers (ReduceLROnPlateau, ExponentialLR, LambdaLR, OneCycleLR), and loss functions. Among the strategies we have tested:

1. Training with the **Cross Entropy (CE)**
2. Training with the **Weighted Cross Entropy (WCE)**
3. Training with the **KappaLoss**
4. Training with the **WCE** $+ \lambda \cdot$ **KappaLoss**
5. Training with the $\alpha \cdot$**WCE** $+ (1 - \alpha) \cdot$**KappaLoss**
6. Training with the **WCE** and finetuning with the **KappaLoss**

With:

- $KappaLoss = 1$ - Quadratic Weighted Kappa Score. The Quadratic Weighted Kappa Score was computed using the TorchMetrics framework.
- λ is a balancing weight factor and α is a weight factor that decreases linearly with the number of epochs.

We faced convergence issue during the training of our approaches with the KappaLoss alone or in combination with WCE in our objective function. We were not able to optimize the weights of our models correctly. The best results were obtained with the CE.

CE is used as a loss function, and Kappa is used to select the best checkpoint. The default data augmentation and Adam optimizer with an initial learning rate of 10^{-4} (weight_decay=10^{-4}) were used to train different backbones. The learning rate decay strategy is ExponentialLR with gamma equal to 0.99. The training epoch is 1000 and the batch size is 4.

Once our baselines were trained, we proceeded to improve them. We used the pseudo-labeling technique. Pseudo-labeling is a process that consists in adding confidently predicted test data to the training data and retraining the models [7]. Through pseudo-label learning, we improved the classification performance of the model after obtaining a good baseline model. Figure 2 illustrates our proposed pseudo-label learning method.

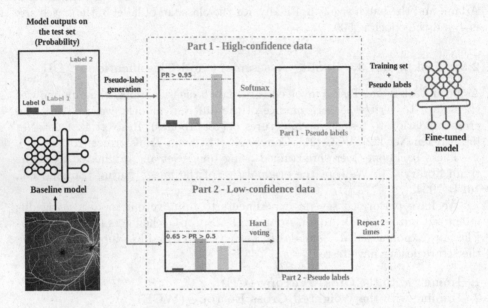

Fig. 2. Our pseudo-label learning method.

According to the probabilities generated by the baseline model on the test set, we have separated the result into two chunks: the first is the high confidence data, and the second is the low confidence data. As with traditional pseudo-label learning methods [7], we considered data with predicted classification probabilities greater than 0.95 to be high-confidence and passed their probability through the Softmax function to determine the pseudo-label.

Repeated training on hard-to-classify samples, like those involved in the Online Hard Example Mining method [20], can enhance the model's performance. Therefore, we used data with probabilities between 0.5 and 0.65, and generated pseudo-labels based on five backbones through hard voting. In order to ensure the accuracy of pseudo-labeling of low-confidence data, the following filtering rules were applied.

1. The baseline model results as pseudo-labels if at least two of the other four backbones have the same result.
2. The results of the other four backbones as pseudo-labels if they are consistent.

Those remaining cases cannot be pseudo-labeled, so their data will not be used.

Since our baseline model performed well, the high-confidence pseudo-labels are more accurate, adding additional data to the training set can improve model robustness.

By hard voting and developing filtering rules, we made the pseudo-labels as accurate as possible for low-confidence data. With the help of this part of the data, the model can make more accurate judgments on data without distinctive features (with probabilities of between 0.65 and 0.9) and improve confidence levels on the remaining uncertain data.

Both pseudo-labels were added to the training set, and the baseline model was fine-tuned. Due to the small size of the second part of the pseudo-label data, we repeated it twice. Furthermore, the results can be further enhanced by iterative pseudo-label learning.

3 Results and Discussion

3.1 Evaluation Metrics

For the first task (segmentation), the Dice similarity coefficient (DSC) and the intersection of union (IoU) are used to evaluate the performance of the segmentation methods.

The Dice coefficient (also known as the Sørensen-Dice coefficient and F1 score) is defined as two times the area of the intersection of A and B, divided by the sum of the areas of A and B. The IOU (Intersection Over Union, also known as the Jaccard Index) is defined as the intersection's area divided by the union's area.

For tasks 2 (quality assessment) and 3 (classification), the organizers propose to use the quadratic weighted kappa and the area under the curve (AUC) to evaluate the performance of classification methods. In particular, they used the macro-AUC One VS One to calculate the AUC value.

Table 3. Segmentation results

Version	Label 1	Label 2	Label 3
V1 - nnU-Net	0.2278	0.6515	0.4621
V2 - V-Net	0.4079	0.6515	0.5259
V3 - V-Net + Classifier	0.4079	0.6515	0.5566

3.2   Segmentation Results

The segmentation task dataset contains 109 images, and each image can have one or more labels. In the training set, there are 86 images that contain the label (1), 106 images that possess the label (2), and only 35 images with the label (3).

We first used the nnU-net method to test the segmentation of the three labels. By analyzing the first results, we noticed that the amount of data for the label (2) is relatively large, so nnU-Net obtained good performances (Dice = 0.6515). However, for labels (1) and (3), the model performed poorly. Therefore, we chose V-Net as the backbone of our binary segmentation for both these labels.

We fine-tuned the V-Net by modifying the loss function, the optimizer, the scheduler, the learning rate, etc. During these experiments, we divided the training dataset into two subsets: 90% for training and 10% for validation. After several runs, we got good results; label (1) has a dice of 0.4079.

After the first tests, we noticed that the distribution of the 65 images in the test set differed from the training set; more than 40 images in the V-Net prediction result contained the label (3). Therefore, we improved our proposed solution by adding a classifier that can detect whether the image contains the label (3) or not.

After the training, the model classified 28 images in class (1), and the rest in class (0), which means that only 28 images contain the label (3), and the others are over-segmented. The label (3) was removed from all the images classified as (0) by the counter-classifier. This step improved the segmentation results for label (3) from 0.5259 to 0.5566.

3.3   Quality Assessment Results

The second task of the DRAC22 challenge concerns the image quality assessment. The training set consists of 665 data, distributed as follows: 97 images belonging to class (0) Poor quality level, 518 belonging to class (1) Good quality level, and 50 images belonging to class (2) Excellent quality level. The first observation is that the data set is very unbalanced.

We used a nested five-fold cross-validation strategy to evaluate different models and backbones. We respected the distribution of the training set in the generation of the folds. For each split, we used four folds for training and validation (20% random as the validation set and 80% as the training set) and one fold for testing. A suitable checkpoint is selected from the validation set, and the final performance of the model is calculated using the test set (one-fold data). This strategy avoids overfitting and provides a more accurate representation of the model's classification performance.

Table 4 summarizes the different backbones used and the results obtained. The model performed poorly on the Fold 4 dataset. On the other hand, the different backbones perform well on Folds 0, 1, 2, and 3. Therefore, we chose the two most optimally performing checkpoints for each fold and tested them on the test set. Their results are shown in Table 5. Val Kappa refers to the one-fold test results in Table 4, whereas Test Kappa refers to the DRAC22 test dataset results. The V5 - VGG19-Fold2 and V2 - VGG16-Fold0 checkpoints performed the best out of the eight selected checkpoints. In order to optimize the use of the training data, we selected the checkpoints V5 - VGG19-Fold2 and performed fine-tuning (random 20% validation set) on the entire training dataset, which

Table 4. Kappa results for different backbones on a one-fold test set.

Backbone	Fold 0	Fold 1	Fold 2	Fold 3	Fold 4	Mean
Resnet50	0.7179	0.8247	0.7502	0.8409	0.5585	0.7384
Resnet101	0.7540	0.7857	0.7515	0.8060	0.6199	0.7434
Resnet152	0.8135	0.7600	0.8537	0.8559	0.5291	0.7624
Resnet200d	0.5488	0.7910	0.7889	0.8593	0.5486	0.7073
Densenet121	0.8525	0.7814	0.8299	0.7966	0.5274	0.7575
Densenet161	0.8357	0.7666	0.8761	0.8358	0.5569	0.7742
Densenet169	0.7942	0.7547	0.8680	0.7908	0.4468	0.7309
Densenet201	0.7980	0.8028	0.8289	0.8563	0.5132	0.7598
Efficientnet_b0	0.7048	0.7617	0.7859	0.8246	0.5367	0.7227
Efficientnet_b1	0.8267	0.7503	0.8349	0.7958	0.5759	0.7567
Efficientnet_b2	0.8406	0.8039	0.8434	0.8563	0.5311	0.7750
Efficientnet_b3	0.7710	0.7787	0.8821	0.7768	0.4801	0.7377
Efficientnet_b4	0.8367	0.8202	0.8855	0.8468	0.5254	0.7829
VGG11	0.8420	0.7730	0.8795	0.8606	0.6199	0.7950
VGG16	0.8592	0.8231	0.8551	0.9168	0.5077	0.7923
VGG19	0.8787	0.8042	0.8933	0.8632	0.3192	0.7517
VGG13	0.8302	0.8409	0.8504	0.8233	0.6704	0.8030
Convnext_base	0.8343	0.8118	0.8531	0.8516	0.5451	0.7792
Swin_base_patch4_window7_224	0.7560	0.7181	0.7236	0.7677	0.2665	0.6464

Table 5. Performance of different checkpoints on the test set.

Check-points	Val Kappa	Test Kappa
V1 - VGG19-Fold0	0.8787	0.7034
V2 - VGG16-Fold0	0.8592	0.7202
V3 - VGG13-Fold1	0.8409	0.7045
V4 - VGG16-Fold1	0.8231	0.6991
V5 - VGG19-Fold2	0.8933	0.7333
V6 - Efficientnet_b4-Fold2	0.8855	0.7184
V7 - VGG16-Fold3	0.9168	0.6987
V8 - VGG19-Fold3	0.8632	0.7154
V9 - VGG19-Finetune	0.8548	0.7447

gave us the baseline model V9 - VGG19-Finetune that has a kappa value of 0.7447 on the test set.

We generated pseudo-labels for each image on the test set based on the baseline model using the classification probabilities. As illustrated in Fig. 2, we treated prediction results for data with probabilities greater than 0.95 (part 1) as pseudo-labels. In cases where the classification probability was between 0.5 and 0.65 (part 2), pseudo-labels were generated using a hard-voting method.

Using all the training sets, we retrained the best-performing four backbones based on the mean of each checkpoint in Table 4. Together with the baseline model, these four checkpoints were hard-voted. Table 6 shows the hard voting results for some of the low-confidence data. The pseudo-labels for 14 of the 20 (6 Unsure) part 2 data were generated and added to the training set by repeating them twice.

Table 6. Hard voting to produce pseudo labels.

Image	VGG19	VGG13	VGG16	VGG11	Efficientnet_b4	Pseudo label
952.png	2	1	2	1	1	Unsure
986.png	2	2	1	2	1	2
1220.png	1	2	1	1	2	1
1283.png	2	1	1	1	1	1
901.png	2	1	1	2	1	Unsure
897.png	2	2	2	2	2	2
611.png	1	1	2	1	1	1

Table 7 illustrates the effectiveness of our pseudo-label learning method. Following pseudo-label learning with the data from part 1, the Kappa value increased from 0.7447 to 0.7484. As a result of pseudo-label learning using data from part 2, the Kappa value increased from 0.7447 to 0.7513. The results indicate that both parts of the data are essential for improving the classification performance. The baseline model's classification performance was significantly improved with a Kappa value of 0.7589 when both parts of the data were used for pseudo-label'learning.

In addition to the baseline model VGG19, we also performed pseudo-label learning on VGG16, resulting in a Kappa of 0.7547. Thus, our Kappa improved to 0.7662 after performing the model ensemble on VGG19 and VGG16.

This result was used to update the pseudo-labels in the second round. There are 399 images from part 1, and 7 from part 2 (4 pseudo labels, 3 unsure). As a result of the first round of pseudo-label learning, the model was also enhanced. The updated pseudo-labels were used to finetune VGG19, resulting in a kappa value of 0.7803.

Table 7. Pseudo-label Ablation study

Method	Val Kappa	Test Kappa
VGG19 (Baseline)	0.8548	0.7447
VGG19-Pseudo-label part 1	0.9458	0.7484
VGG19-Pseudo-label part 2	0.8830	0.7513
VGG19-Pseudo-label part 1 + part 2	0.8733	0.7589

3.4   DR Grading Results

The objective of this third task is DR grading. The training dataset groups 611 patients into three grades: label (0) - Normal - (329 images), label (1) - NPDR - (212 images), and finally label (2) - PDR - (70 images).

To process this task, we followed the same steps as task 2. Firstly, we performed five-fold cross-validation and selected the eight best-performing checkpoints on the test set. The two most appropriate checkpoints were selected based on the kappa in the test set: V1 - DenseNet121-Fold1 and V2 - Efficientnet b3-Fold3. The baseline model was then fine-tuned using all the training sets: V3 - DenseNet121-Finetune and V4 - Efficientnet_b3-Finetune. Secondly, we generated pseudo-labels based on the classification results of V3 - DenseNet121-Finetune. In the first round of pseudo-label learning, there were 266 images for part 1, and 20 images for part 2 (15 pseudo labels, 5 unsure). We then performed the first round of pseudo-label learning for DenseNet121 and Efficientnet_b3. Next, we performed a model ensemble (Kappa = 0.8628) and obtained new pseudo-labels. In the second round of pseudo-label learning, there were 332 images for part 1 and 12 images for part 2 (7 pseudo labels, 5 unsure). After the second round of pseudo-label learning, we performed a model ensemble on DenseNet121 and Efficientnet_b3 and obtained the final Kappa 0.8761.

Table 8. Classification results for different steps

Chick-point	Val Kappa	Test Kappa
V1 - DenseNet121-Fold1	0.8275	0.8100
V2 - Efficientnet_b3-Fold3	0.8776	0.8069
V3 - DenseNet121-Finetune	0.9335	0.8370
V4 - Efficientnet_b3-Finetune	0.9728	0.8239
V5 - DenseNet121-First round	0.9542	0.8499
V6 - Efficientnet_b3-First round	0.9112	0.8545
V7 - DenseNet121-Second round	0.9674	0.8520
V8 - Efficientnet_b3-Second round	0.9558	0.8662

4   Conclusion

In this article, we summarized our participation in the DRAC22 challenge. We showed that despite the efficiency of the nnU-Net method in the segmentation task, it does not always give good results, especially when the data set is relatively small. However, the fine-tuning of the V-Net model allowed us to overcome this limitation by obtaining better results for both labels (1) and (3).

During the test phase, we noticed many images containing label (3), which was inconsistent with the distribution of the training set. Adding an independent model for the binary classification of the images (either containing label (3) or not) before the segmentation improved our result for this label.

For tasks 2 and 3, the pseudo-labeling allowed us to improve our models progressively. Indeed, training baselines, using them to label the test set, and then keeping the labeled images with high confidence allows the model to have more data in the second training round. This iterative process allowed our models to perform better.

We have also shown that ensembles of models can generate good performance and allow us to label low-confident data. Indeed, the ensembles can overcome bias and variance from different architectures. Models help each other and cancel each other's errors, resulting in higher accuracy.

As for our work in progress, we are combining tasks for better segmentation and classification. We think that using the segmentation results could guide the classifier of task 3. Also, we noticed that when the image quality is poor, this image is still segmented in task 1, giving us segmented regions that should not exist. So we believe that using the models of task 2 before the segmentation could improve the performance.

References

1. Atwany, M.Z., Sahyoun, A.H., Yaqub, M.: Deep learning techniques for diabetic retinopathy classification: a survey. IEEE Access **10**, 28642–28655 (2022). https://doi.org/10.1109/ACCESS.2022.3157632
2. Dai, L., et al.:A deep learning system for detecting diabetic retinopathy across the disease spectrum. Nature Commun. **12**(1) (Dec 2021). https://doi.org/10.1038/s41467-021-23458-5
3. De Carlo, T.E., Romano, A., Waheed, N.K., Duker, J.S.: A review of optical coherence tomography angiography (octa). Int. J. Retina Vitreous **1**(1), 5 (2015). https://doi.org/10.1186/s40942-015-0005-8
4. He, K., Zhang, X., Ren, S., Sun, J.: Deep residual learning for image recognition (2015). 10.48550/ARXIV.1512.03385
5. Huang, G., Liu, Z., van der Maaten, L., Weinberger, K.Q.: Densely connected convolutional networks (2016). 10.48550/ARXIV.1608.06993, https://arxiv.org/abs/1608.06993
6. Isensee, F., et al.: nnu-net: Self-adapting framework for u-net-based medical image segmentation (2018). 10.48550/ARXIV.1809.10486, https://arxiv.org/abs/1809.10486
7. Lee, D.H., et al.: Pseudo-label: The simple and efficient semi-supervised learning method for deep neural networks. In: Workshop on Challenges in Representation Learning, ICML. vol. 3, p. 896 (2013)
8. Li, Y., et al.: Multimodal information fusion for glaucoma and diabetic retinopathy classification. In: Antony, B., Fu, H., Lee, C.S., MacGillivray, T., Xu, Y., Zheng, Y. (eds.) Ophthalmic Medical Image Analysis, pp. 53–62. Springer International Publishing, Cham (2022). https://doi.org/10.1007/978-3-031-16525-2_6
9. Liu, R., et al.: Deepdrid: Diabetic retinopathy-grading and image quality estimation challenge. Patterns **3**(6), 100512 (2022)
10. Liu, Z., et al: Swin transformer: Hierarchical vision transformer using shifted windows (2021). 10.48550/ARXIV.2103.14030, https://arxiv.org/abs/2103.14030
11. Liu, Z., Mao, H., Wu, C.Y., Feichtenhofer, C., Darrell, T., Xie, S.: A convnet for the 2020s (2022). 10.48550/ARXIV.2201.03545, https://arxiv.org/abs/2201.03545
12. Milletari, F., Navab, N., Ahmadi, S.A.: V-net: Fully convolutional neural networks for volumetric medical image segmentation (2016)
13. Quellec, G., et al.: 3-d style transfer between structure and flow channels in oct angiography. Invest. Ophthalmol. Vis. Sci. **63**(7), F0259-2989 (2022). https://doi.org/10.1109/ACCESS.2022.3157632
14. Quellec, G., Al Hajj, H., Lamard, M., Conze, P.H., Massin, P., Cochener, B.: Explain: Explanatory artificial intelligence for diabetic retinopathy diagnosis. Med. Image Anal. **72**, 102118 (2021) 10.1016/j.media.2021.102118, https://www.sciencedirect.com/science/article/pii/S136184152100164X
15. Quellec, G., Charrière, K., Boudi, Y., Cochener, B., Lamard, M.: Deep image mining for diabetic retinopathy screening. Med. Image Anal. **39**, 178–193 (2017) 10.1016/j.media.2017.04.012, https://www.sciencedirect.com/science/article/pii/S136184151730066X
16. Ronneberger, O., Fischer, P., Brox, T.: U-net: Convolutional networks for biomedical image segmentation (2015)
17. Russell, J., Shi, Y., Hinkle, J., Scott, N., Fan, K., Lyu, C., Gregori, G., Rosenfeld, P.: Longitudinal wide-field swept-source oct angiography of neovascularization in proliferative diabetic retinopathy after panretinal photocoagulation. ophthalmol retina. Retina **3**(4), 350–361 (2019). https://doi.org/10.1016/j.oret.2018.11.008

18. Schaal, K.B., Munk, M.R., Wyssmueller, I., Berger, L.E., Zinkernagel, M.S., Wolf, S.: Vascular abnormalities in diabetic retinopathy assessed with swept-source optical coherence tomography angiography widefield imaging. Retina **39**(1), 79–87 (2019). https://doi.org/10.1097/IAE.0000000000001938

19. Sheng, B., Chen, X., Li, T., Ma, T., Yang, Y., Bi, L., Zhang, X.: An overview of artificial intelligence in diabetic retinopathy and other ocular diseases. Frontiers in Public Health 10 (2022). https://doi.org/10.3389/fpubh.2022.971943,https://www.frontiersin.org/articles/10.3389/fpubh.2022.971943

20. Shrivastava, A., Gupta, A., Girshick, R.: Training region-based object detectors with online hard example mining. In: Proceedings of the IEEE Conference on Computer Vision and Pattern Recognition, pp. 761–769 (2016)

21. Simonyan, K., Zisserman, A.: Very deep convolutional networks for large-scale image recognition (2014). 10.48550/ARXIV.1409.1556, https://arxiv.org/abs/1409.1556

22. Tan, M., Le, Q.V.: Efficientnet: Rethinking model scaling for convolutional neural networks (2019). 10.48550/ARXIV.1905.11946, https://arxiv.org/abs/1905.11946

23. Tian, M., Wolf, S., Munk, M.R., Schaal, K.B.: Evaluation of different swept'source optical coherence tomography angiography (ss-octa) slabs for the detection of features of diabetic retinopathy. Acta Ophthalmologica **98**(4), e416–e420 (2020) 10.1111/aos.14299, https://onlinelibrary.wiley.com/doi/abs/10.1111/aos.14299

24. Wightman, R.: Pytorch image models. https://github.com/rwightman/pytorch-image-models (2019). https://doi.org/10.5281/zenodo.4414861

25. Zeghlache, R., et al.: Detection of diabetic retinopathy using longitudinal self supervised learning. In: Antony, B., Fu, H., Lee, C.S., MacGillivray, T., Xu, Y., Zheng, Y. (eds.) Ophthalmic Medical Image Analysis, pp. 43–52. Springer International Publishing, Cham (2022). https://doi.org/10.1007/978-3-031-16525-2_5

26. Zhang, Q., Rezaei, K.A., Saraf, S.S., Chu, Z., Wang, F., Wang, R.K.: Ultra-wide optical coherence tomography angiography in diabetic retinopathy. Quant. Imaging Med. Surgery **8**(8) (2018), https://qims.amegroups.com/article/view/21249

Data Augmentation by Fourier Transformation for Class-Imbalance: Application to Medical Image Quality Assessment

Zhicheng Wu[ID], Yanbin Chen[ID], Xuru Zhang[ID], and Liqin Huang[(⊠)][ID]

College of Physics and Information Engineering, Fuzhou University, Fuzhou, China
hlq@fzu.edu.cn

Abstract. Diabetic retinopathy (DR) is a common ocular disease in diabetic patients. In DR analysis, doctors first need to select excellent-quality images of ultra wide optical coherence tomography imaging (UW-OCTA). Only high-quality images can be used for lesion segmentation and proliferative diabetic retinopathy (PDR) detection. In practical applications, UW-OCTA has a small number of images with poor quality, so the dataset constructed from UW-OCTA faces the problem of class-imbalance. In this work, we employ data enhancement strategy and develop a loss function to alleviate class-imbalance. Specifically, we apply Fourier Transformation to the poor quality data with limited numbers, thus expanding this category data. We also utilize characteristics of class-imbalance to improve the cross-entropy loss by weighting. This method is evaluated on DRAC2022 dataset, we achieved Quaratic Weight Kappa of 0.7647 and AUC of 0.8458, respectively.

Keywords: Deep learning · Image quality assessment · Fourier transformation · Class-aware weighted loss

1 Introduction

Globally, the proportion of people with diabetes is significant and in a phase of rapid growth. According to the International Diabetes Federation (IDF), the number of adults with diabetes worldwide would reach 537 million (10.5%) in 2021. Diabetic retinopathy (DR) is one of the common complications of diabetes. Diabetes can cause damage to retinal microvasculature, which in turn affects vision and may lead to blindness. Ultrawide optical coherence tomography angiography (UW-OCTA) is helpful for retinopathy analysis [1,2]. Based on a study by UK BioBank, more than 25% of retinal images are of insufficient quality for accurate lesion diagnosis [3]. As shown in Fig. 1, in the UW-OCTA image, the image with good quality has clear texture, while the image with poor quality has problems such as image blur and artifacts. Lower-quality images

---

Z. Wu and Y. Chen—The two authors have equal contributions to the paper.

B. Sheng and M. Aubreville (Eds.): MIDOG 2022/DRAC 2022, LNCS 13597, pp. 161–169, 2023.
https://doi.org/10.1007/978-3-031-33658-4_15

interference doctors' diagnosis of diseases such as lesions segmentation and PDR detection. Therefore, it is necessary to develop an automatic retinal image quality assessment system to save ophthalmologists' assessment time.

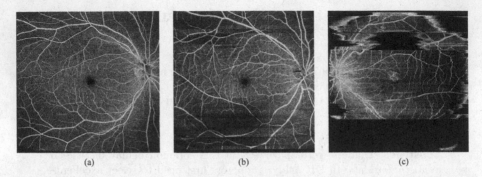

Fig. 1. Image quality level: (a) Excellent quality level; (b) Good quality level; (c) Poor quality level.

Image Quality Assessment (IQA) plays an important role in image acquisition technology, image reconstruction and image post-processing algorithms. Dai et al. [4] developed a transfer learning-assisted multi-task network for assessing retinal image quality, retinopathy and DR grades. At present, the image quality evaluation of the imaging department mainly depends on the subjective evaluation of the radiologist [5,6]. The most commonly used method in the image quality control management is still manual investigation, supervision and reporting. Saha et al. [7] proposed an automatic retinal image quality assessment system, which achieved 100% accuracy when the dataset only had two categories of "accepting" and "not accepting", but after adding the third category of "ambiguous" images, accurate rate dropped significantly. Niemeijer et al. [8] introduced an image structure clustering method to extract a compact representation of retinal structure to determine the image quality level. Tajbakhsh et al. [9] adopted different augmentation methods to synthesize poor quality images to provide sufficient data for network training, the results shows a significant improvement in performance.

Most of the datasets used in previous work have only two categories. There are few datasets with three categories, and the data distribution of three categories is uniform. UW-OCTA dataset provided from the DRAC2022 Challenge, as shown in Table 1, contains three different image quality levels: Poor quality level (0), Good quality level (1), Excellent quality level (2). The dataset contains 50 images of the poor category, 97 images of the good category, 518 images of the excellent category. It is obvious that the images of the three categories are seriously unbalanced.

For class-imbalance, it means that the number of one class is much larger or smaller than the other classes. To address the class-imbalance problem, people usually preprocess the data. At present, undersampling and oversampling methods are widely applied in data preprocessing. The former focuses on reducing the

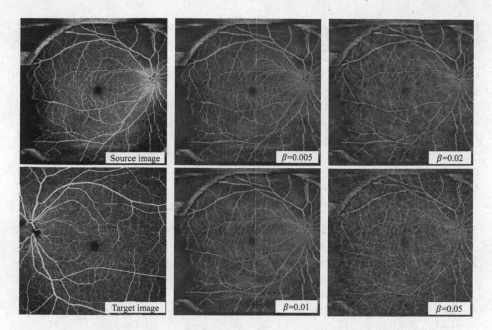

Fig. 2. Effect of the size of the α: increasing α will generate different style images but it introduces noise and artifacts.

number of majority classes, while the latter focuses on amplifying the number of minority classes. For undersampling, Alejo et al. [10] introduced ENN, for those samples of the majority class, to remove the junction of the class edge or even the minority class cluster. For oversampling, Sáez et al. [11] adopted SMOTE to enlarge the minority class size and proposed adding a noise filter method to remove some noisy examples from both majority and minority classes. Different instance selection alone can yield satisfactory results even if the undersampled dataset is still class-imbalance as long as a certain proportion of noisy data is filtered from the majority class [12,13]. Previous work has tended to propose a sub-network to solve this problem, but this strategy complicates the overall network. Recently, the Fourier transformation method has also proven to be very effective For data enhancement [14]. Inspired from the above work, we adopt the Fourier Transform, a concise and efficient method, to alleviate class-imbalance. We know that the phase component of the Fourier spectrum preserves the high-level semantics of the original signal, while the magnitude component contains

Table 1. Statistics of different Image Quality Levels on the UW-OCTA dataset of the 2022 diabetic retinopathy analysis challenge.

	Excellent quality level	Good quality level	Poor quality level
Number	518	97	50

low-level statistics. High-level semantics can be understood as the structural information of the image, and low-level statistical information can be understood as the style information of the image. We propose to augment the training data by distorting the amplitude information while keeping the phase information unchanged.

In order to maintain the category balance and avoid model laziness (the full prediction of Excellent quality will achieve the highest accuracy rate), we employ Fourier Transformation to perform data enhancement for the images of poor quality level. The dataset is expanded while maintaining the balance of categories, and the loss function is improved by weighting, which greatly improves the generalization ability of the model.

This paper has the following contributions:

1. We propose an automatic image quality assessment framework and evaluate it on the UW-OCTA dataset of the 2022 diabetic retinopathy analysis challenge.
2. We employ Fourier Transformation to enhance images with poor quality, alleviate class-imbalance and expand training samples at the same time, which improves the generalization ability of the model.
3. We modify the cross-entropy loss to make the model pay more attention to the small number of samples.

2   Proposed Method

It has been proved that deep learning technology can achieve desirable performance in retinal image quality assessment [15]. Nowadays, most methods are only applicable to the situation where the number of images in different categories is relatively close. However, in practical applications, there is a problem of class-imbalance. In this paper, we hope that we can use data enhancement strategies to alleviate the problem of class- imbalance, which can not only make data diversified, but also make the model more Robustness. The overall framework of the network is shown in Fig. 3. For the image of Poor quality level, we randomly sample the image of Excellent quality level, obtain its amplitude spectrum using Fourier Transformation, and apply the amplitude spectrum to enhance it. The labels of the enhanced images are still 0 (Because the enhanced image still belongs to the category of poor quality). We attempt to make the network learn more about the poor quality features and avoid making mistakes. Then we add the enhanced images to the dataset and obtain a balanced dataset as input. The network output quality features represent vectors. We use class-aware weighted loss to guide network training to further improve model generalization. It is worth noting that we have also done image enhancement operations including random rotation, random scaling, random clipping, and random flipping, except the above data enhancement strategies. When we enhance the dataset, the number of the three categories is balanced.

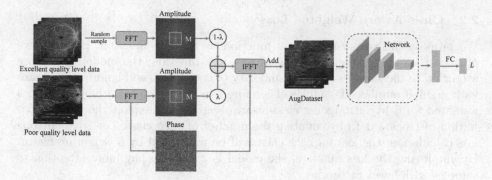

Fig. 3. schematic of the workflow of the proposed image quality assessment framework.

2.1 Data Augmentation via Fourier Transformation

Fourier Transformation is a very important algorithm in the field of digital image processing. For image I of poor quality level, we randomly select an image I' from the image in the excellent quality level in the training set, perform Fourier Transformation on I to obtain amplitude spectrum A and phase spectrum P, and then perform the same Fourier Transformation on I' to obtain its amplitude spectrum A'. We introduce the amplitude information of image I' to enhance I:

$$A_f = (1 - \lambda)A * (1 - M) + \lambda A' * M \tag{1}$$

where A_f is the amplitude spectrum after fusing A', λ is used to adjust the ratio of the amplitude spectrum A and A', M is a binary mask, used to control the range of the amplitude spectrum to be exchanged, for image I, we take the amplitude low frequency information, for the image I', we take its high-frequency information.

Then, we adopt the inverse Fourier Transformation to transform the fused amplitude spectrum into the spatial domain, thus obtaining the enhanced image Z:

$$Z = \mathcal{F}^{-1}(\mathcal{A}_f, \mathcal{P}) \tag{2}$$

Choice of α: We define the size of M divided by the size of I as α. As we can see from the Eq. 1, when $\alpha = 0$, the enhanced image is the same as the source image. On the other hand, when $\alpha = 1.0$, the amplitude spectrum of the source image will be replaced by the amplitude spectrum of another image. Figure 2 illustrates the effect of α. We find that as α increases, the image produces different style changes, but it brings more noise and artifacts. When $\alpha \leq 0.01$, the enhanced image maintains the same structure as the source image and has no Introduce too many external factors. To achieve our goal of enhancing the image, so we set $\alpha \leq 0.01$

2.2   Class-Aware Weighted Loss

We propose the cross-entropy loss function as the loss function. However, we observe that only using the cross-entropy loss function, the model pays more attention to the samples with the majority of categories, and ignores the classes with a small number of samples. Therefore, we modify the loss function into a weighted form by introducing class-aware weights. We exploit the class distribution of labeled data by counting the number of each class. For cross-entropy loss calculation, the loss for each class will be multiplied by a weighting factor. By employing the loss function, the model is guided to pay more attention to samples with fewer categories.

$$w_i = \left(\frac{\max \{n_j\}_{j=0}^C}{n_i} \right)^\beta, \quad n_i = \frac{N_i}{\sum_{j=0}^C N_j}, \quad i = 0, \ldots, C \qquad (3)$$

where C is the number of categories, N_i represents the number of the ith category, and n_i represents the proportion of the ith category. In the experiment, the exponential term β is empirically set to 1/4.

3   Experimental Results

3.1   Dataset

DRAC2022 Challenge provides standardized ultra wide (sweep frequency) optical coherence tomography angiography (UW-OCTA) data set. The dataset includes images and tags for automatic image quality assessment, lesion segmentation, and DR grading. We choose IQA dataset for training and validation. There are 665 images in the training set and 438 images in the test set, The instrument used in this challenge was a SS-OCTA system (VG200D, SVision Imaging, Ltd., Luoyang, Henan, China), works near 1050nm and features a combination of industry-leading specifications including ultrafast scan speed of 200,000 AScans per second.

3.2   Implementation Details

We implemented the model on Pytorch 1.12.1 and trained it by using NVidia GTX 1080Ti GPU with 16GB RAM on the Ubuntu20.04 system. We utilized four networks, Resnet34 [16], ResNet50 [16], Densenet121 [17] and EfficientNet-b2 [18], and adopted the weights trained on ImageNet [19]. At the end of network, the full connected layer is added as a three classifier, and the parameters of the full connected layer are initialized randomly. We leveraged AdamW to optimize the network with the weight decay of 0.1, the learning rate of 0.0001, and the batch size of 32. We trained 50 epochs on the overall architecture. In addition, we adopt random rotation, random scaling, random clipping, and random flipping as data enhancement strategies. For excellent quality level images, we do not employ enhancement Strategies, for good quality level images, we use 4 enhancement strategies, for poor quality level images, we randomly sample 5 amplitude spectra of excellent quality level to enhance, also using 4 enhancement strategies. So our enhanced dataset comes with 518 images of excellent quality level, 485 images of good quality level, and 500 images of poor quality level. We divide the enhanced dataset into training set and validation set according to the ratio of 8:2, and we also adopt the strategy of 5-fold cross validation. For evaluation metrics, quadratic weighted kappa and Area Under Curve (AUC) are used to evaluate the performance of the classification methods.

3.3 Result

In order to find the most suitable image quality assessment network, we compared four networks. EfficientNet-b2 was finally determined. Secondly, in order to verify the effectiveness of our method, we also conducted a number of ablation experiments to verify the effectiveness of Fourier Transformation in data enhancement and the improved loss function.

Table 2. Performances of different methods on the UW-OCTA dataset of the 2022 diabetic retinopathy analysis challenge.

Method	Quaratic Weight Kappa	AUC
ResNet34	0.5811	0.8497
ResNet50	0.6190	0.8120
DenseNet121	0.6956	**0.8647**
EfficientNet-b2	**0.7647**	0.8458

Table 3. Ablation study on the UW-OCTA dataset of the 2022 diabetic retinopathy analysis challenge. CA Loss means Class-aware weighted loss.

Fourier	CA Loss	Quaratic Weight Kappa	AUC
		0.6627	0.8672
✓		0.7262	**0.8791**
	✓	0.6986	0.8669
✓	✓	**0.7647**	0.8458

The performance of our method on several different networks is shown in Table 2. Although the AUC of EfficientNet-b2 is slightly smaller than that of Densenet121, its kappa index is better than other networks, because the kappa index can better represent the goodness in unbalanced data, so employing the EfficientNet-b2 network structure can be more effective. The problem of balance and the classification effect are also relatively good. Table 3 shows a series of ablation experiments, all of which are based on EfficientNet-b2. It can be found that using only the classification network for this task, even a good classification network will not be able to perform fully on this task (e.g. the network will be biased towards predicting "2", a category with a high number of images, but not really learning the quality features of the images). Experiments show that after the Fourier Transformation is applied to data enhancement, the kappa value increases significantly. The results indicates that Fourier Transformation alleviate the class-imbalance problem and improving the generalization of the network. Meanwhile, the performance of classification is better after modifying the loss funcion.

4   Conclusion

In this paper, in order to solve the problem of imbalance data set categories, we apply Fourier Transformation and modify the loss function to solve this problem. The idea is relatively simple, but the performance of classification is greatly improved. We validated the proposed method on DRAC2022 dataset and achieved satisfactory results. The experimental results demonstrate that the data augmentation operation by Fourier Transformation can robustly classify the different categories of images when facing class-imbalance. In the future, we will develop more strategies to improve our method, such as labeling the image with poor quality, reducing the sample with good quality, modifying the network structure, and proposing a more targeted loss function.

Acknowledgements. Research supported by National Natural Science Foundation of China (62271149) and Fujian Provincial Natural Science Foundation project (2021J02019).

References

1. Russell, J.F., et al.: Longitudinal wide-field swept-source OCT angiography of neovascularization in proliferative diabetic retinopathy after panretinal photocoagulation. Ophthalmol. Retina **3**(4), 350–361 (2019)
2. Zhang, Q., Rezaei, K.A., Saraf, S.S., Chu, Z., Wang, F., Wang, R.K.: Ultra-wide optical coherence tomography angiography in diabetic retinopathy. Quant. Imaging Med. Surg. **8**(8), 743 (2018)
3. Fu, H., et al.: Evaluation of retinal image quality assessment networks in different color-spaces. In: Shen, D., et al. (eds.) MICCAI 2019. LNCS, vol. 11764, pp. 48–56. Springer, Cham (2019). https://doi.org/10.1007/978-3-030-32239-7_6
4. Dai, L., et al.: A deep learning system for detecting diabetic retinopathy across the disease spectrum. Nat. Commun. **12**(1), 1–11 (2021)
5. Liu, R., et al.: DeepDRiD: diabetic retinopathy-grading and image quality estimation challenge. Patterns **3**(6), 100512 (2022)
6. Sheng, B., et al.: An overview of artificial intelligence in diabetic retinopathy and other ocular diseases. Front. Public Health **10**, 971943 (2022)

7. Saha, S.K., Fernando, B., Cuadros, J., Xiao, D., Kanagasingam, Y.: Automated quality assessment of colour fundus images for diabetic retinopathy screening in telemedicine. J. Digit. Imaging **31**(6), 869–878 (2018). https://doi.org/10.1007/s10278-018-0084-9
8. Niemeijer, M., Abramoff, M.D., van Ginneken, B.: Image structure clustering for image quality verification of color retina images in diabetic retinopathy screening. Med. Image Anal. **10**(6), 888–898 (2006)
9. Tajbakhsh, N., et al.: Convolutional neural networks for medical image analysis: full training or fine tuning? IEEE Trans. Med. Imaging **35**(5), 1299–1312 (2016)
10. Alejo, R., Sotoca, J.M., Valdovinos, R.M., Toribio, P.: Edited nearest neighbor rule for improving neural networks classifications. In: Zhang, L., Lu, B.-L., Kwok, J. (eds.) ISNN 2010. LNCS, vol. 6063, pp. 303–310. Springer, Heidelberg (2010). https://doi.org/10.1007/978-3-642-13278-0_39
11. Sáez, J.A., Luengo, J., Stefanowski, J., Herrera, F.: SMOTE-IPF: addressing the noisy and borderline examples problem in imbalanced classification by a re-sampling method with filtering. Inf. Sci. **291**, 184–203 (2015)
12. Lin, W.-C., Tsai, C.-F., Ya-Han, H., Jhang, J.-S.: Clustering-based undersampling in class-imbalanced data. Inf. Sci. **409**, 17–26 (2017)
13. Tsai, C.-F., Lin, W.-C., Ya-Han, H., Yao, G.-T.: Under-sampling class imbalanced datasets by combining clustering analysis and instance selection. Inf. Sci. **477**, 47–54 (2019)
14. Yao, H., Hu, X., Li, X.: Enhancing pseudo label quality for semi-supervised domain-generalized medical image segmentation. arXiv preprint arXiv:2201.08657 (2022)
15. Yu, F.L., Sun, J., Li, A., Cheng, J., Wan, C., Liu, J.: Image quality classification for DR screening using deep learning. In: 2017 39th Annual International Conference of the IEEE Engineering in Medicine and Biology Society (EMBC), pp. 664–667. IEEE (2017)
16. He, K., Zhang, X., Ren, S., Sun, J.: Deep residual learning for image recognition. In: Proceedings of the IEEE Conference on Computer Vision and Pattern Recognition, pp. 770–778 (2016)
17. Huang, G., Liu, Z., Van Der Maaten, L., Weinberger, K.Q.: Densely connected convolutional networks. In: Proceedings of the IEEE Conference on Computer Vision and Pattern Recognition, pp. 4700–4708 (2017)
18. Tan, M., Le, Q.: EfficientNet: rethinking model scaling for convolutional neural networks. In: International Conference on Machine Learning, pp. 6105–6114. PMLR (2019)
19. Deng, J., Dong, W., Socher, R., Li, L.-J., Li, K., Fei-Fei, L.: ImageNet: a large-scale hierarchical image database. In: 2009 IEEE Conference on Computer Vision and Pattern Recognition, pp. 248–255. IEEE (2009)

Automatic Image Quality Assessment and DR Grading Method Based on Convolutional Neural Network

Wen Zhang(ID), Hao Chen(ID), Daisong Li(ID), and Shaohua Zheng(✉)(ID)

College of Physics and Information Engineering, Fuzhou University, Fuzhou 350108, FJ, China
sunphen@fzu.edu.cn

Abstract. Diabetic retinopathy (DR) is a common ocular complication in diabetic patients and is a major cause of blindness in the population. DR often leads to progressive changes in the structure of the vascular system and causes abnormalities. In the process of DR analysis, the image quality needs to be evaluated first, and images with better imaging quality are selected, followed by value-added proliferative diabetic retinopathy (PDR) detection. Therefore, in this paper, the MixNet classification network was first used for image quality assessment (IQA), and then the ResNet50-CMBA network was used for DR grading of images, and both networks were combined with a k-fold cross-validation strategy. We evaluated our method at the 2022 Diabetic Retinopathy Analysis Challenge (DRAC), where image quality was evaluated on 1103 ultra-wide optical coherence tomography angiography (UW-OCTA) images and DR grading was detected on 997 UW-OCTA images. Our method achieved a Quadratic Weight Kappa of 0.7547 and 0.8010 in the test cases, respectively.

Keywords: Diabetic retinopathy grading · Image quality assessment · Ultra-wide optical coherence tomography angiograph

1 Introduction

Diabetic retinopathy (DR) is a severe and widely spread eye disease. It is the commonest cause of blindness in the working-age population of developed countries [1]. Diabetes often causes progressive changes in the vascular structure of the fundus and the resulting abnormalities. DR is diagnosed by visually inspecting retinal fundus images for the presence of retinal lesions, such as microaneurysms (MAs), intraretinal microvascular abnormalities (IRMAs), nonperfusion areas and neovascularization. The detection of these lesions is critical to the diagnosis of DR. The ultra-wide optical coherence tomography angiography (UW-OCTA) images of the healthy retina, the non-proliferative diabetic retinopathy (NPDR) and the proliferative diabetic retinopathy (PDR) (see Fig. 1). Blurred images can disguise lesions and the image does not show the correct part of the retina, so

B. Sheng and M. Aubreville (Eds.): MIDOG 2022/DRAC 2022, LNCS 13597, pp. 170–177, 2023.
https://doi.org/10.1007/978-3-031-33658-4_16

that a diseased eye could be mistakenly graded as normal. Therefore, the quality of the image is equally important for the diagnosis of DR, and the quality of the image is divided into three levels (see Fig. 2). Thus, we need to evaluate the image quality of UW-OCTA and select the images with better imaging quality and perform DR analysis.

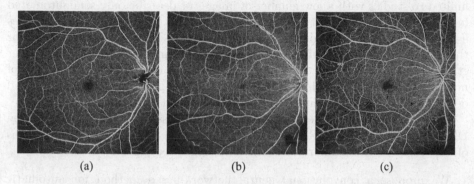

(a)   (b)   (c)

Fig. 1. DR Grading: (a) Healthy Retina; (b) NPDR; (c) PDR.

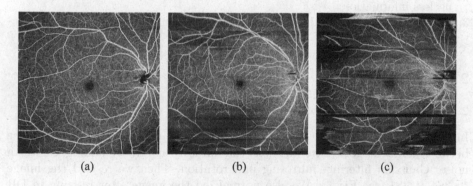

(a)   (b)   (c)

Fig. 2. Image quality: (a) Excellent; (b) Good; (c) Poor.

Artificial intelligence (AI) is a broad term encompassing multiple components of which machine learning (ML) and deep learning (DL) are the two major components [2]. Deep learning has become a research hotspot in the medical field, and its development has effectively promoted the progress of DR research. Dai et al. developed a deep learning system called DeepDR, which can detect early to advanced stages of diabetic retinopathy [3]. Liu et al. described a challenge named "Diabetic Retinopathy (DR)—Grading and Image Quality Estimation Challenge" to develop deep learning models for DR image assessment and grading [4]. Traditional DR grading diagnosis is an invasive fundus imaging and cannot be used in patients with allergies, pregnancy, or poor liver or kidney function. Ultra-wide OCTA can detect changes in DR neovascularization non-invasively and is therefore an important imaging tool to help ophthalmologists diagnose PDR. However, there are no works that can perform automated DR analysis with UW-OCTA.

Currently, published works on automated retinal image clarity assessment are limited to studies with a low number of images or describe only semiautomated methods [5–7]. Image quality assessment (IQA) is not very accurate and requires a lot of time and effort. For DR grading, Wu et al. [1] proposed a method for the automatic detection of microaneurysms in retinal fundus images. Maher et al. already evaluated a decision support system for automatic screening of non-proliferative DR [8]. However, despite all these previous works, automated detection of DR still remains a field for improvement [9].

In this paper, we first perform image quality assessment by MixNet and then DR grading of images by ResNet50-CMBA network. Overall, the contributions of our work can be summarized in the following three aspects:

1. We propose a convolutional neural network-based method for automatic image quality assessment and DR grading and evaluate it on the UW-OCTA dataset of the 2022 Diabetic Retinopathy Analysis Challenge (DRAC 2022).
2. We incorporate attention mechanisms [10] in the network to obtain more detailed information about the desired target of attention and suppress other useless information.
3. We use k-fold cross validation to solve the problem of limited data and effectively avoid overfitting.

2   Methods

2.1   Data Augmentation

In the data augmentation stage, in order to slove the problem of imbalance between data categories, we augment the dataset, mainly by adding Gaussian noise, Gaussian filtering, mirroring and rotation. Then we resized the image to 384×384 (see Fig. 3). We also normalized the image after resizing in DR grading.

2.2   Image Quality Assessment

For IQA, we trained a MixNet with 5-fold cross validated method (see Fig. 4). MixNet is a lightweight network that mixes different kernel sizes in a single convolutional operation based on the AutoML search space. The 5-fold cross validation first divides the dataset into 5 mutually exclusive subsets of the same size at random, and each time selects 4 copies as the training set and the remaining 1 copy as the test set. Each subset can only be selected once as a test set. A total of 5 rounds of training are performed, and then the models and parameters that the optimal loss function for the 5 rounds are saved. In the testing phase,

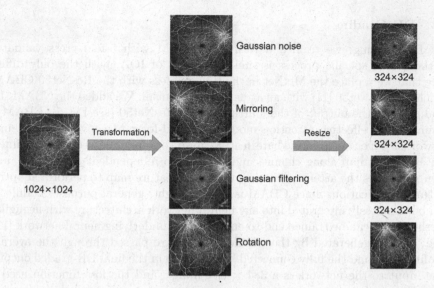

Fig. 3. Data augmentation.

we added up the prediction probability values of the 5 best models and take the category with the highest probability value as the final category prediction. The input to the network is a 384 × 384 image and the output is classified into three levels of image quality by linear transformation: excellent, good and poor. The loss function used is CrossEntropyLoss:

$$CEloss(g, p) = - \sum_{i=1}^{C} g_i \log (p_i) \qquad (1)$$

where C denotes the number of classes, g_i is ground truth, p_i is prediction.

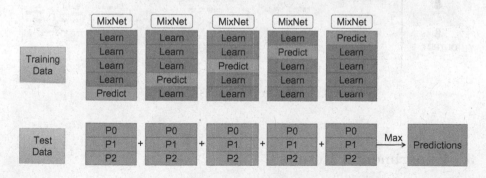

Fig. 4. 5-fold cross-validated MixNet network.

2.3   DR Grading

For DR grading, we trained a ResNet50-CBAM with 5-fold cross validated method. The specific process is similar to that of IQA, with the only difference that we replace the MixNet in the IQA process with the ResNet50-CBAM, which is a ResNet50 [11] with an attention mechanism. We added the CBAM [10] module after the output of the fourth stage of ResNet50 (see Fig. 5). CBAM is a simple and effective attention module for feed-forward convolutional neural networks. Given an intermediate feature map, our module sequentially infers the attention map along channel and space, two independent dimensions, and then multiplies the attention map by the input feature map to perform adaptive feature modification. Since CBAM is a lightweight, general-purpose module, it can be seamlessly integrated into any neural network architecture with negligible overhead and can be trained end-to-end with the underlying neural network [11]. The features generated by the CBAM module are passed through the average pooling layer and the fully connected layer to obtain the final DR-graded output. The input to the network is a 384×384 image, and the loss function used is CrossEntropyLoss.

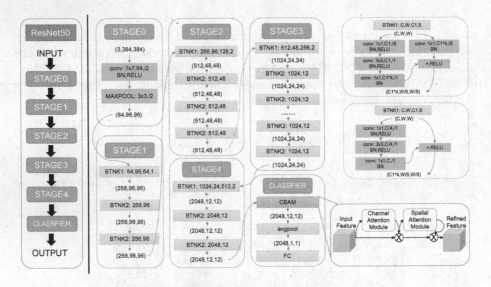

Fig. 5. ResNet50-CBAM.

3   Experimental Results

3.1   Dataset

The DRAC2022 provides a standardized ultra-wide UW-OCTA dataset. The dataset includes images and labels for automated image quality assessment, lesion segmentation, and DR grading. We selected the dataset of IQA and DR grading for training and validation. The IQA dataset contains 665 training images and 1103 test images. After data augmentation, there are 1303 training images. We divide 261 of the training images as validation images. The DR grading dataset contains 611 training images and 997 test images. After data augmentation, there are 1809 training images. We divide 362 of the training images as validation images.

3.2 Implementation Details

As an experimental setting, we choose PyTorch to implement our model and use an NVIDIA GeForce RTX 3060 GPU for training. The input size of the networks is 384×384 with a batch size of 8. In our model, we set the epoch to 40, and the initial learning rate to 1×10^{-5}. We use the Quadratic Weight Kappa and area under the curve (AUC) to evaluate the performance of the classification method.

4 Result

To find the optimal solution of the method, we compared the two networks. In addition to exploring the attention mechanism and the advantages of k-fold cross validation, we performed ablation experiments with the same dataset and data parameters. We utilized the evaluation metrics Quadratic Weighted Kappa and area under the curve (AUC) to validate our method on the DRAC2022 dataset.

Table 1. Quadratic Weight Kappa and AUC for image quality assessment.

Method	Quadratic Weight Kappa	AUC
ResNet50	0.5646	0.8341
ResNet50-CBAM	0.6565	0.8595
MixNet	0.7481	0.8771
5-fold MixNet	**0.7547**	**0.8862**

Table 1 shows the experimental results of IQA. From Table 1, MixNet is more suitable than ResNet50 for the task of IQA, reaching 0.7481 in Quadratic Weight Kappa, and the network combined with the CBAM module performs better in both Quadratic Weight Kappa and AUC. The MixNet with 5-fold cross validation also improved the Quadratic Weight Kappa by 0.0129 over that of the single MixNet.

Table 2 shows the experimental results of DR grading. From Table 2, ResNet50 is more suitable than MixNet for the task of IQA, reaching 0.7481 in Quadratic Weight Kappa, and the network combined with the CBAM module improved by 0.0193 in Quadratic Weight Kappa. The use of class activation

Table 2. Quadratic Weight Kappa and AUC for DR grading.

Method	Quadratic Weight Kappa	AUC
MixNet	0.7441	0.8337
ResNet50	0.7754	0.8614
ResNet50-CBAM	0.7947	0.8701
5-fold ResNet50-CBAM	**0.8010**	**0.8885**

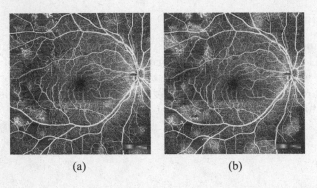

(a)   (b)

Fig. 6. Class activation maps: (a) ResNet50; (b) ResNet50-CBAM.

maps shows where the weight or center of gravity of the model was during the training process (see Fig. 6). Analyzing the images it can be concluded that the weight or center of gravity of the model with the CBAM module added is better focused on the location where the lesions occur in the fundus vessels. ResNet50-CBAM with 5-fold cross validation also achieve better results than the single ResNet50-CBAM, reaching 0.8010 in Quadratic Weight Kappa.

5   Conclusion

In this work, we developed an automatic image quality assessment and DR grading method based on convolutional neural networks. We used MixNet to implement automatic image quality assessment and then designed ResNet50-CBAM to implement DR classification detection. We included an attention mechanism in ResNet50-CBAM to obtain more detailed information about the target to be focused on while suppressing other irrelevant information. For both tasks, we used 5-fold cross validation to solve the problem of limited data. We tested it on the DRAC2022 UW-OCTA dataset. The experimental results show that the method achieves good results in both image quality assessment and DR grading.

References

1. Wu, B., Zhu, W., Shi, F., Zhu, S., Chen, X.: Automatic detection of microaneurysms in retinal fundus images. Comput. Med. Imaging Graph. **55**, 106–112 (2017)

2. Sheng, B., et al.: An overview of artificial intelligence in diabetic retinopathy and other ocular diseases. Front. Public Health **10**, 971943 (2022). https://doi.org/10.3389/fpubh.2022.971943

3. Dai, L., Wu, L., Li, H., et al.: A deep learning system for detecting diabetic retinopathy across the disease spectrum. Nat. Commun. **12**(1), 1–11 (2021)

4. Liu, R., Wang, X., Wu, Q., et al.: DeepDRiD: diabetic retinopathy-grading and image quality estimation challenge. Patterns **3**(6), 100512 (2022)

5. Lalonde, M., Gagnon, L., Boucher, M.: Automatic visual quality assessment in optical fundus images. In: Proceedings of Vision Interface, Ottawa, Ontario, Canada, pp. 259–264 (2001). https://www.cipprs.org/vi2001/schedulefifinal.html

6. Lee, S.C., Wang, Y.: Automatic retinal image quality assessment and enhancement. In: Proceedings of SPIE Medical Imaging Processing, Washington, DC, vol. 3661, pp. 1581–1590. SPIE (1999)

7. Choong, Y.F., Rakebrandt, F., North, R.V., Morgan, J.E.: Acutance, an objective measure of retinal nerve fibre image clarity. Br. J. Ophthalmol. **87**, 322–326 (2003)

8. Maher, R., Kayte, S., Panchal, D., Sathe, P., Meldhe, S.: A decision support system for automatic screening of non-proliferative diabetic retinopathy. Int. J. Emerg. Res. Manag. Technol. **4**(10), 18–24 (2015)

9. Maher, R., Kayte, S., Dhopeshwarkar, D.M.: Review of automated detection for diabetes retinopathy using fundus images. Int. J. Adv. Res. Comput. Sci. Softw. Eng. **5**(3), 1129–1136 (2015)

10. Woo, S., Park, J., Lee, J.-Y., Kweon, I.S.: CBAM: convolutional block attention module. In: Ferrari, V., Hebert, M., Sminchisescu, C., Weiss, Y. (eds.) ECCV 2018. LNCS, vol. 11211, pp. 3–19. Springer, Cham (2018). https://doi.org/10.1007/978-3-030-01234-2_1

11. He, K., Zhang, X., Ren, S., et al.: Deep residual learning for image recognition. In: Proceedings of the IEEE Conference on Computer Vision and Pattern Recognition, pp. 770–778 (2016)

A Transfer Learning Based Model Ensemble Method for Image Quality Assessment and Diabetic Retinopathy Grading

Xiaochao Yan ®, Zhaopei Li ®, Jianhui Wen, and Lin Pan (✉)

College of Physics and Information Engineering, Fuzhou University,
Fuzhou 350108, China
panlin@fzu.edu.cn

Abstract. Diabetic retinopathy (DR) is a chronic complication of diabetes that damages the retina and is one of the leading causes of blindness. In the process of diabetic retinopathy analysis, it is necessary to first assess the quality of images and select the images with better imaging quality. Then DR analysis, such as DR grading, is performed. Therefore, it is crucial to implement a flexible and robust method to achieve automatic image quality assessment and DR grading. In deep learning, due to the high complexity, weak individual differences, and noise interference of ultra-wide optical coherence tomography angiography (UW-OCTA) images, individual classification networks have not been able to achieve satisfactory accuracy on such tasks and do not generalize well. Therefore, in this work, we use multiple models ensemble methods, by ensemble different baseline networks of RegNet and EfficientNetV2, which can simply and significantly improve the prediction accuracy and robustness. A transfer learning based solution is proposed for the problem of insufficient diabetic image data for retinopathy. After doing feature enhancement on the images, the UW-OCTA image task will be fine-tuned by combining the network pre-trained with ImageNet data. our method achieves a quadratic weighted kappa of 0.778 and AUC of 0.887 in image quality assessment (IQA) and 0.807 kappa and AUC of 0.875 in diabetic retinopathy grading.

Keywords: Image Quality Assessment · Diabetic Retinopathy Grading · Model Ensemble · Transfer Learning

1 Introduction

Diabetic retinopathy (DR) is a chronic complication of diabetes that damages the retina. Patients with diabetic retinopathy are 25 times more likely to go blind than healthy individuals, making diabetic retinopathy one of the leading causes of human blindness. DR usually causes progressive changes in vascular structure and leads to abnormalities [1]. DR screening mainly relies on color fundus images,

B. Sheng and M. Aubreville (Eds.): MIDOG 2022/DRAC 2022, LNCS 13597, pp. 178–185, 2023.
https://doi.org/10.1007/978-3-031-33658-4_17

and retinal images are collected from patients by fundus image acquisition equipment, and finally ophthalmologists perform manual analysis and lesion screening. Because of the small variability between different grades in diabetic retinal images, and the difference in experience of professional ophthalmologists attending the clinic, misdiagnosis and missed diagnoses may occur. There has been some work using fundus images for DR diagnosis [2, 3]. However, it is difficult to detect early or small neovascular lesions with fundus photography. ultra-wide optical coherence tomography angiography imaging (UW-OCTA) enhances the imaging of microvascular morphology through changes in dynamic blood flow, is fast and does not require the use of contrast media and is an important imaging modality to help ophthalmologists diagnose PDR. There are no works that can perform automated DR analysis using UW-OCTA. During DR analysis, the image quality of UW-OCTA first needs to be evaluated and images with better imaging quality are selected. Then DR analysis, such as lesion segmentation and PDR detection, is performed. Therefore, it is crucial to build flexible and robust models for automatic image quality assessment, lesion segmentation, and PDR detection.

Nowadays, due to the advancement of artificial intelligence, the detection of diabetic retinopathy and image quality assessment by the automated system has become a trend in modern medical diagnosis [4]. Kang et al. [5] first proposed the use of convolutional neural network model for the image quality assessment (IQA) task and proved its effectiveness experimentally. Mahapatra et al. [6] proposed a CNN network containing convolutional layer, maximum pooling operation, full connected layer CNN network, which is based on the human visual system and can classify fundus image quality but requires a large data sample for training, otherwise, overfitting will occur. Yu et al. [7] applied transfer learning to the image quality classification task, but only fine-tuned the ALexNet network. For lesion classification, Gulshan et al. [8] used the Inception-V3 model framework and employed more than 120,000 diabetic retinal fundus images to detect lesions, and the method achieved good performance due to a sufficient amount of data as well as expert screening. Sharma et al. [9] proposed a deep CNN architecture for the classification of non-DR and DR images. Zhou et al. [10] used a multi-module architecture to learn high-resolution fundus images and trained them to predict labels with classification and regression. Wang et al. [11] used an attention mechanism to focus on suspicious lesion regions and accurately predicted the level of lesions using the whole image and high-resolution suspicious lesion patches, but the method required pixel-level annotation. Philippe et al. [12] developed a method to predict the level of lesions by GAN artificially synthesized retinal images to expand the dataset and improve the ability to train DR grading models. There are researchers have recently addressed these tasks in combination. Dai et al. [13] developed a deep learning system called DeepDR for assessing retinal image quality, retinal lesions and DR grading and obtained high sensitivity and accuracy.

Currently, deep learning has achieved good classification results for diabetic retinopathy, but these are based on large data volumes or accurate annotation. And the assessment of image quality is still a task that needs to be improved.

Therefore, in this work we use a migration learning-based solution, using network weights based on ImageNet-1K pre-training, and set the network classification layer output to the corresponding class, fine-tuned with the UW-OCTA dataset. Using the model ensemble method, the prediction probabilities of multiple models are averaged for the final prediction, making the model's prediction less sensitive to the details of the training data, the choice of the training scheme, and the chance of a single training run.

Our main contributions to this work are as follows:

1. We propose a new image classification method for image quality assessment and diabetic retinopathy classification, using multiple models different baseline networks for integration.
2. To address the problem of insufficient data, the network is fine-tuned using weights based on ImageNet-1K pre-training to reduce the network training cost.
3. We evaluate our proposed method on the DARC2022 challenge dataset. The validity can be well demonstrated.

2   Method

In this section, we describe the general framework of the proposed method in detail. The details about the method are described as follows.

In the preprocessing stage, as shown in Fig. 1, the proposed method includes the following steps:

1) Resize the original grayscale image to 384 × 384,
2) Do histogram equalization of the original image,
3) Do binary change of the grayscale image using Gaussian weighted adaptive thresholding,
4) Stitch the original image, the equalized image, and the binary image into a three-channel image.

For the image quality assessment task, the preprocessing step is performed only up to step 1.

An overview of our proposed framework is shown in Fig. 2. By training multiple models, the models are selected for the EfficientNetV2 [14] and RegNet [15] networks that perform better in the ImageNet dataset for classification. By training different baselines of EfficientNetV2 and RegNet. EfficientNetV2-S, RegNetY-800MF and RegNetY-3.2GF are selected for task 2 network, and EfficientNetV2-S, RegNetY-800MF, RegNetY-1.6GF and RegNetY-3.2GF are selected for task 3 baseline network. Each baseline model has its own characteristics and each model predicts a different error. The model ensemble is done by averaging the predicted outcome probabilities.

To reduce the training cost as well as to address the insufficient amount of training data, transfer learning is used to achieve support for the new task by starting training the neural network model with the pre-trained model as

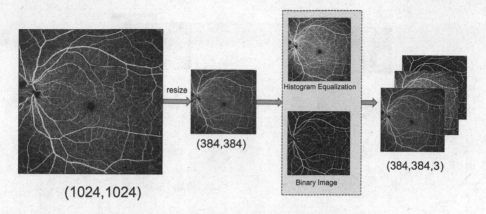

Fig. 1. Image pre-processing steps.

a checkpoint. The weights pre-trained by the natural image dataset ImageNet-1K are migrated to the medical image, and the full weights are loaded and retrained because of the great difference in style between the natural image and the medical image. Due to the constraints of the pre-trained weights make the training speed and the prediction accuracy greatly improved compared to the random initialization of parameters.

To category imbalance of the dataset for the two tasks of IQA and DR grading, we deal with the problem by the following two methods. First, the category imbalance can be mitigated to some extent by using data augmentation. The input images are randomly flipped horizontally and vertically, brightness, contrast, and hue are randomly varied. The data are normalized according to the mean and standard deviation of the training data set. Second, the training strategy incorporates cost-sensitivity by calculating the proportion of each category in the dataset, and then taking the inverse of the ratio as the weight, adding the weight to the loss function so that the error loss for a few categories generates a stronger update of the network weights. The loss function is as follows:

$$WCEloss(g,p) = -\sum_{i=1}^{C} wg_i \log{(p_i)} \tag{1}$$

where C denotes the number of classes, w is the weight of each category, g_i is the label, p_i is prediction.

3   Experiments

3.1   Dataset and Evaluation Measures

In this paper, we use the UW-OCTA dataset, which provides more than 600 fundus grayscale images for image quality assessment and DR grading training. The training set for image quality assessment consists of 665 images and the

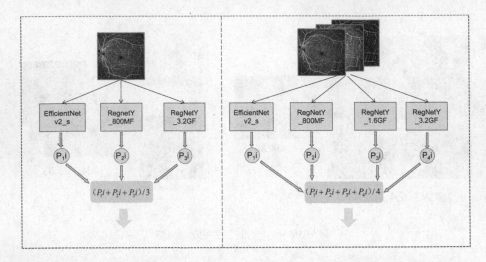

Fig. 2. Overview diagram of the model ensemble, with the method of image quality assessment on the left, ensemble three models, and the method of diabetic retinopathy grading on the right, with the input being a processed fused three-channel image and ensemble four models for prediction

corresponding labels in the CSV file. The dataset contains three different image quality grades as shown in Fig. 3: poor quality level (0), good quality level (1), and excellent quality level (2). Similarly the DR grading training set consists of 611 images and the corresponding labels in the CSV file. It contains three different diabetic retinopathy grades: Normal (0), NPDR (1), and PDR (2). The evaluation metrics include two accuracy metrics: the performance of the classification method using quadratic weighted kappa and the area under curve (AUC).

3.2 Implementation Details

The development environments and requirements are presented in Table 1.

Table 1. Development environments and requirements.

Windows/Ubuntu version	Ubuntu 20.04.4 LTS
CPU	Intel(R) Core(TM) i9-12900K
GPU (number and type)	Two NVIDIA Tesla P40
CUDA version	11.7
Programming language	Python 3.8
Deep learning framework	Pytorch (1.12.0,torchvision 0.13.0)

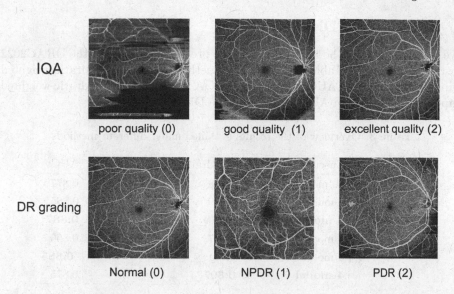

IQA

poor quality (0)   good quality (1)   excellent quality (2)

DR grading

Normal (0)   NPDR (1)   PDR (2)

Fig. 3. Overview of UW-OCTA dataset

In our training process, the data are divided into training and validation sets in the ratio of 0.8 to 0.2. The following enhancements are applied to the data: 1). Input images are randomly (50 % probability) flipped horizontally and vertically. 2) Random variations are done for brightness, contrast and hue. 3) Data are normalized based on the mean and standard deviation of the training dataset. The weights of the loss function is done to calculate the proportion of each class in the dataset, then the weight is taken as the inverse of the ratio, and the weight is added to the loss function making the error loss generated for a few classes to update the network weights more strongly. IQA task weights are set to (0.6, 0.3, 0.1), DR grading weights are (0.15, 0.25, 0.6). The details of our training scheme are shown in Table 2.

Table 2. Training protocols.

Input size	(384,384)
Network pre-training weights	Pre-trained by IMAGENET1K
Batch size	16
Total epochs	40
Optimizer	AdamW (weightdecay = 0.1)
Initial learning rate (lr)	1e−3
Learning rate scheduler	LinearWarmupCosineAnnealingLR
Loss Function	Weighted Cross Entropy

4   Results and Discussion

We evaluated our method by presenting the prediction results on the DRAC2022 test set. As shown in Table 3, our proposed method achieves a quadratic weighted kappa of 0.778 and an AUC of 0.887 on the IQA test set, and a quadratic weighted kappa of 0.807 and an AUC of 0.875 on the DR graded test set.

Table 3. Overview of the results obtained under different methods.

	Ensemble methods	Quadratic Weighted Kappa	AUC
IQA	10 model	0.748	0.877
	5 model	0.769	0.885
	3 model	**0.778**	**0.887**
DR grading	5 model	0.737	0.860
	3 model	0.791	**0.885**
	4 model	**0.807**	0.875

For the IQA task, we initially trained 10 networks, namely ResNet34, ResNet34, ResNet50, Densenet121, EfficientNet-b2, EfficientNet-b3, EfficientNet-V2, RegNetY-800MF, RegNetY-1.6GF, and RegNetY-3.2GF. But due to the poor performance of some networks such as ResNet on the validation set, ResNet, Densenet and EfficientNet-b3 were no longer used in the subsequent model ensemble, and only five models were used for integration, and the metrics were found to have improved. The reason for this phenomenon is that the model integration uses a simple weighted average, which is easy to predict wrongly for networks with low performance like ResNet, and the number of networks accounts for too much, which leads to the final results being biased towards the wrong results of networks like ResNet. Out of this situation, we finally choose 3 networks that perform best on the validation set for ensemble, and the final result performs best on the test set. As the DR classification, we considered it as a task of the same nature as IQA, so we migrated the model weights trained in the IQA task to the DR classification task, and only retrained the classification layer, and found that several baselines of RegNet with a small number of parameters performed well for this task, and finally used them all for the ensemble test. Compared to the IQA task, the method does not achieve a higher level of performance on the DR task. This phenomenon may be due to the fact that a single network has reached a performance bottleneck for DR hierarchy tasks, and integrating multiple models does not improve the performance significantly.

5   Conclusion

In this work, we proposed a model ensemble method for image quality assessment and diabetic retinopathy grading, which can solve the problem of insufficient training data by transferring some classical classification networks to medical image tasks. And for the category imbalance problem, is overcome by data enhancement and weighted cross entropy. The model ensemble approach can solve the problem of poor generalization of individual networks. The best performing model among all training models on the validation set was selected for ensemble. We validated the effectiveness of the method on the UW-OCTA dataset.

References

1. Tian, M., Wolf, S., Munk, M.R., Schaal, K.B.: Evaluation of different Swept'Source optical coherence tomography angiography (SS-OCTA) slabs for the detection of features of diabetic retinopathy. Acta ophthalmol. **98**(1), e416–e420 (2019)
2. Dai, L., et al.: A deep learning system for detecting diabetic retinopathy across the disease spectrum. Nat. Commun. **12**(1), 3242 (2021)
3. Lyu, X., Jajal, P., Tahir, M.Z., Zhang, S.: Fractal dimension of retinal vasculature as an image quality metric for automated fundus image analysis systems. Sci. Rep. **12**(1), 1–13 (2022)
4. Sheng, B., et al.: An overview of artificial intelligence in diabetic retinopathy and other ocular diseases. Front. Public Health **10** (2022)
5. Le, K., Peng, Y., Yi, L., Doermann, D.: Convolutional neural networks for no-reference image quality assessment. In: 2014 IEEE Conference on Computer Vision and Pattern Recognition (2014)
6. Mahapatra, D., Roy, P.K., Sedai, S., Garnavi, R.: A CNN based neurobiology inspired approach for retinal image quality assessment. In: International Conference of the IEEE Engineering in Medicine & Biology Society (2016)
7. Yu, F., Sun, J., Li, A., Cheng, J., Liu, J.: Image quality classification for DR screening using deep learning. In: Engineering in Medicine & Biology Society (2017)
8. Gulshan, V.: Development and validation of a deep learning algorithm for detection of diabetic retinopathy in retinal fundus photographs. Jama **316**(22), 2402–2410 (2016)
9. Sunil, S., Saumil, M., Anupam, S.: An intelligible deep convolution neural network based approach for classification of diabetic retinopathy. Bio-Algorithms Med-Syst. **14** (2018)
10. Zhou, K., et al.: Multi-cell multi-task convolutional neural networks for diabetic retinopathy grading. In: 2018 40th Annual International Conference of the IEEE Engineering in Medicine and Biology Society (EMBC) (2018)
11. Wang, Z., Yin, Y., Shi, J., Fang, W., Li, H., Wang, X.: Zoom-in-Net: deep mining lesions for diabetic retinopathy detection. In: Descoteaux, M., Maier-Hein, L., Franz, A., Jannin, P., Collins, D.L., Duchesne, S. (eds.) MICCAI 2017. LNCS, vol. 10435, pp. 267–275. Springer, Cham (2017). https://doi.org/10.1007/978-3-319-66179-7_31
12. Burlina, P.M., Joshi, N., Pacheco, K.D., Liu, T.A., Bressler, N.M.: Assessment of deep generative models for high-resolution synthetic retinal image generation of age-related macular degeneration. JAMA Ophthalmol. **137**, 258–264 (2019)
13. Dai, L., et al.: A deep learning system for detecting diabetic retinopathy across the disease spectrum. Nat. Commun. **12**(1), 1–11 (2021)
14. Tan, M., Le, Q.: EfficientNetV2: smaller models and faster training. In: International Conference on Machine Learning, pp. 10096–10106 (2021)
15. Radosavovic, I., Kosaraju, R.P., Girshick, R., He, K., Dollár, P.: Designing network design spaces. In: 2020 IEEE/CVF Conference on Computer Vision and Pattern Recognition (CVPR) (2020)

Automatic Diabetic Retinopathy Lesion Segmentation in UW-OCTA Images Using Transfer Learning

Farhana Sultana[1]([✉])(iD), Abu Sufian[1](iD), and Paramartha Dutta[2](iD)

[1] Department of Computer Science, University of Gour Banga, West Bengal, India
sfarhana128@gmail.com
[2] Department of CSS, Visva-Bharati University, West Bengal, India

Abstract. Regular retinal screening and timely treatment are the only ways to avoid vision loss due to Diabetic Retinopathy (DR). However, the insufficiency of ophthalmologists or optometrists makes DR screening and treatment programs challenging for the growing global diabetic population. Computer-aided automatic DR screening and detection systems will be a more sustainable approach to deal with this situation. The Diabetic Retinopathy Analysis Challenge 2022 (DRAC22) in association with the 25th International Conference on Medical Image Computing and Computer Assisted Intervention 2022 (MICCAI 2022) created the opportunity for researchers worldwide to work on automatic DR diagnosis procedure on UW-OCTA images of the retina. As automatic segmentation of different DR lesions is the first and most crucial step in the DR screening procedure, we addressed the task of "Segmentation of Diabetic Retinopathy Lesions" among three different tasks of the challenge. We used the transfer learning technique to automatically segment lesions from the retinal images. The chosen pre-trained deep learning model is trained, validated, and tested on the DRAC22 segmentation dataset. It showed a mean Dice Similarity Coefficient (DSC) of 32.06% and a mean Intersection over Union (IoU) of 22.05% on the test dataset during the challenge submission. Some variations in the training procedure lead the model's performance to a mean DSC of 43.36 % and a mean IoU of 31.03% on the test dataset during post-challenge submission. The link to the repository of code is: https://github:com/Sufianlab/FS_AS_DRAC22

Keywords: Convolutional Neural Network · Deep Learning · Diabetic Retinaopathy · Medical Image Segmentation · Retinal Image Analysis · Transfer Learning

1 Introduction

A person with diabetes sometimes suffers from vision-related problems called Diabetic Retinopathy (DR). Severe DR most frequently leads to blindness too. According to statistics of the International Diabetes Federation (IDF) [1], at

B. Sheng and M. Aubreville (Eds.): MIDOG 2022/DRAC 2022, LNCS 13597, pp. 186–194, 2023.
https://doi.org/10.1007/978-3-031-33658-4_18

present, 537 million adults (20–79 years) are affected with diabetes and it may rise to 643 million by 2030 and 783 million by 2045. The leading cause of vision loss in working-age adults (20–65 years) is Diabetic Retinopathy. Approximately one in ten people with diabetes suffer from threatening eyesight due to this DR [2]. The problem of vision loss due to DR can be avoided with regular screening and treatment. Nevertheless, the insufficiency of ophthalmologists or optometrists makes DR screening and treatment programs challenging for the growing global diabetic population. In this situation, a computer-aided automatic DR screening and detection system will be a more sustainable approach to adopt.

Ophthalmologists or optometrists diagnose diabetic retinopathy by visually inspecting retinal fundus images for the presence of retinal lesions, such as Microaneurysms (MAs), Intraretinal Microvascular Abnormalities (IRMAs), Nonperfusion Areas (NPAs) and neovascularization (NV). Various invasive fundus imaging modalities are there to diagnose retinal lesions. But invasiveness criteria of fundus imaging may endanger patients with allergies, pregnancy, or poor liver and kidney function. In this situation, the ultra-wide optical coherence tomography angiography (UW- OCTA) imaging modality will be beneficial due to its non-invasiveness behaviour and ability to capture minute details of lesions. There are some works which used retinal fundus images for DR diagnosis [5,10,12]. However, currently, there is no significant work capable of automatic DR analysis using UW-OCTA images. The Diabetic Retinopathy Analysis Challenge 2022 (DRAC22), in association with the 25th International Conference on Medical Image Computing and Computer Assisted Intervention 2022 (MICCAI 2022), created the opportunity for researchers worldwide to work on automatic DR diagnosis procedure on UW-OCTA images [13]. The challenge DRAC22 consists of three tasks: Task 1: Segmentation of Diabetic Retinopathy Lesions, Task 2: Image Quality Assessment, and Task 3: Diabetic Retinopathy Grading. In the DR diagnosis procedure, segmentation of different DR lesions is the first and the most crucial part. Hence, we have chosen the task of "Segmentation of Diabetic Retinopathy Lesions" to participate.

Among various deep learning models, the convolutional neural network (CNN) showed exceptional results in different fundamental works of computer vision such as image classification, object detection, image segmentation and others. On the one hand, the performance of the deep learning models depend on the amount of training data. On the other, accurately annotated medical image segmentation data is scarce. Hence, the nature of data dependency of the deep learning models leads us to train them on a large amount of data. The DRAC22 segmentation dataset contains only 109 original UW-OCTA images for three different types of lesions segmentation: 1. Intraretinal Microvascular Abnormalities (IRMAs), 2. Nonperfusion Areas (NPAs), and 3. Neovascularization (NVs). Training a deep learning model with a small amount of data causes a lack of generalization power. Transfer learning can be a way to handle this problem. In transfer learning, we use an existing pre-trained model for a new task. Various pre-trained model hubs provide deep learning models trained on large image dataset such as ImageNet [6]. In this work, we have used a deep convolutional neural network-based segmentation network named DeepLabV3 [4] to segment DRAC22 dataset. This network uses a pre-trained image classification model called residual network (ResNet50) [7] as its backbone network. ResNet50 is pre-trained on the ImageNet dataset.

We used a pre-trained segmentation model on the DRAC22 segmentation dataset. Hence, we need to fine tune the model on the unknown dataset. The DRAC22 dataset is too small to fine tune a deep model. Therefore, We have augmented the data to handle this data shortage issue. We have discussed our augmentation procedure in Sect. 3.2. A pre-trained model also need specific type of data preprocessing. As DeepLabV3 is pre-trained on the ImageNet dataset, we preprocessed the DRAC22 dataset according to the data preprocessing technique of the ImageNet dataset. We elaborated this in Sect. 3.3.

We have trained, validated and tested our chosen pre-trained model on the DRAC22 segmentation dataset. We have discussed the comparative performance of the model used during challenge submission and post-challenge submission in Sect. 5.

Organization of the paper:
Rest of the paper is organized as follows:

- Section 2 specifies data related information.
- Section 3 briefly describes the methods.
- Section 4 explains the experimental framework.
- Section 5 presents the result and discussion.
- Section 6 concludes the study.

2   Dataset Description

The DRAC22 diabetic retinopathy lesion segmentation dataset contains three different types of lesions: 1. Intraretinal Microvascular abnormalities (IRMAs), 2. Nonperfusion Areas (NPAs) and 3. Neovascularization (NVs). The dataset contains a single folder of 109 original UW-OCTA images and three different folders for three distinct ground truths corresponding to the lesions. For IRMAs, NPAs, and NVs, there are 86, 106, and 35 images respectively. Among the 109 original images, 31 contain all three types of lesions, 56 contain any two types of lesions, and 22 contain only one type of lesion. We have shown the data distribution in Fig. 1. Hence, for 109 original images, a total of 227 ground truth images are given for all three types of lesions. So, there is a dissimilarity in the number of original and ground truth images. To handle this data dissimilarity, we have created three different folders for the original image corresponding to three distinct lesions. If an original image contains all three types of lesions then we put that image in all three folders. In this way, we have created a total of 227 original images for 227 corresponding ground truths. We renamed the original image and corresponding ground truth for our convenience. For example, the original image '82.png' contains all three types of lesions. Hence, we made three copies of the image, renamed it as '82_1.png', '82_2.png', and '82_3.png', and put it into IMRAs, NPAs and NVs folders individually as depicted in Fig. 2.

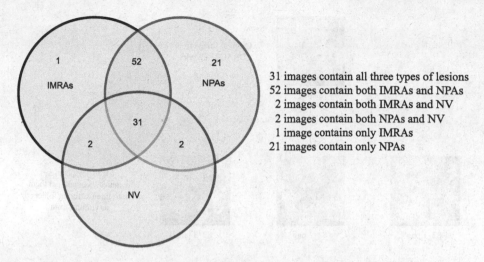

Fig. 1. Original image data distribution corresponding to three different lesions.

31 images contain all three types of lesions
52 images contain both IMRAs and NPAs
2 images contain both IMRAs and NV
2 images contain both NPAs and NV
1 image contains only IMRAs
21 images contain only NPAs

3 Method Description

We have discussed the network used, the data augmentation procedure and the data preprocessing technique in this section.

3.1 Model Description

From the last decade, the enormous success of deep convolutional neural network paves the way for the massive application of different variation of the network in various tasks of computer vision. If we go through the evolution of CNN-based image segmentation models [15], we will find that the backbones of popular segmentation models such as FCN [11], DeepLab [3], etc., are based on basic CNN based image classification networks such as VggNet [14], GoogLeNet [16], ResNet [8], DenseNet [9], SENet [8]. In this work, we have used DeepLabV3 [4] to segment Diabetic Retinopathy (DR) lesions from UW-OCTA images. We chose the pre-trained image classification network ResNet50 as the backbone of this network. The DeepLabV3 model is designed for the PASCAL VOC 2012 dataset consists of 21 distinct segmentation classes. Hence, the last layer of the network has 21 output channels. As the DRAC22 segmentation dataset has a single segmentation class, we have used one output channel in the last classification layer. We also removed the auxiliary classification layer.

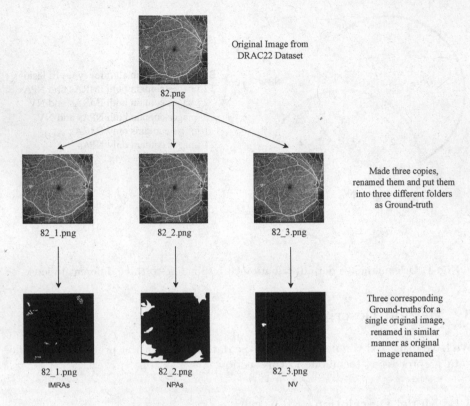

Fig. 2. Procedure for duplicating and renaming the original image and corresponding ground truth.

3.2   Data Augmentation

The thirst for data in deep learning models increases with the model's parameters. Large and complex models, trained on a small amount of data, tend to overfit. In this work, we have used a deep convolutional neural network named DeepLabV3 with backbone ResNet50 as our model for transfer learning. To train this model, we only have a total of 227 original images (renamed from 119 original images) and corresponding 227 ground truths as described in Sect. 2. This data insufficiency lead us to data augmentation. We have described the data augmentation procedure used during challenge submission and post-challenge submission.

Challenge Submission: During challenge submission, we created 512 × 512 non-overlapping image patches from each original and ground truth image. Then, we applied horizontal flip, vertical flip, random flip, rotation, grid distortion, optical distortion, and elastic transforms on each image patch.

Post Challenge Submission: In IMRAs, ground truth contains very small-sized blobs of foreground lesion pixels compared to the image size shown in

Table 1. Minimum, Maximum and Average blob size of foreground lesion pixels in ground truth images.

Lesion Type	Minimum Blob Size	Maximum Blob Size	Average Blob Size
IMRAs	292	68057	8612
NPAs	1584	578403	201239
NV	572	90396	16152

Table 1. Hence, patch creation lead to plenty of empty patches meaning the ground truth patch does not contain any foreground lesion pixels. This downfall of patch creation gave us unsatisfactory performance during challenge submission shown in Table 4. During post-challenge submission, we resized the original images and the ground truths to 512×512. Then, we applied horizontal flip, vertical flip and rotation as augmentation techniques. This procedure enhanced the performance of all three types of lesions. The total number of augmented images per submission is shown in Table 2.

Table 2. Number of original image and number of augmented image per lesion type during challenge submission and post-challenge submission.

Lesion Type	Number of Original Image	Number of Augmented Image	
		Challenge Submission	Post-Challenge Submission
IMRAs	86	2408	4386
NPASs	106	2968	4346
NV	35	980	3535

3.3 Data Preprocessing

ResNet50 [7], the backbone of DeepLabV3 [4], is pretrained on the ImageNet [6] dataset. Hence, to train the DeepLabV3 model using the DRAC22 dataset, we must follow the same data preprocessing techniques used to process the ImageNet dataset. We have listed below the techniques used to preprocess the DRAC22 segmentation dataset.

- We have converted the original gray image (or patch) into three channel RGB image (or patch).
- The input image (or patch) size is 512×512 fulfilling the minimum required image size of 224×224 for the model.
- The range of image pixel value must be in $[0, 1]$. So, we divided each image pixel by 255. Then it is normalized using $mean = [0.485, 0.456, 0.406]$ and $standard\ deviation = [0.229, 0.224, 0.225]$.
- The ground truth images are normalized only by dividing each pixel by 255.

4   Experimental Frame Work

The DeepLabV3 model is trained, validated and tested on the DRAC22 dataset.

Data division: We have divided the dataset into training and validation sets in the following manner as shown in Table 3. We have used the same amount of test data for both submissions.

Table 3. Training, validation and test data used during challenge submission and post-challenge submission.

Submission Type	Lesion Type	Training	Validation	Testing
Challenge Submission	IMRAs	2160	248	
	NPAs	2664	304	65
	NV	882	98	
Post-Challenge Submission	IMRAs	3960	426	
	NPAs	3915	431	65
	NV	3186	349	

Optimization: We trained our model using Adam optimizer with an initial learning rate of 0.0001 and reduced it whenever the validation loss plateaus for five epochs. We trained the model for 100 epochs during challenge submission and 50 epochs for post-challenge submission. We chose a mini-batch size of 4 images to train the model. We used DiceBCE loss as a loss function. This loss function is a combination of the dice loss and the standard binary cross-entropy (BCE) loss. We have used an early stopping criterion during post-challenge submission. In early stopping, if the validation loss increases for five epochs, the training of the model will be stopped.

Cross-Validation: To train the model, we used 10-fold random cross validation. To combat overfitting, we have used 10% of training data to cross validate the model during training. After each epoch, training and validation sets are reorganized automatically.

Evaluation Metrics: We have used Dice Similarity Coefficient (DSC) and Intersection over Union (IoU) as evaluation metric for the model.

5   Result and Discussion

In this section, we analyzed the performance of our model for two distinct submissions: challenge submission and post-challenge submission.

Challenge Submission: During challenge submission, we augmented the DRAC22 diabetic retinopathy lesion segmentation dataset using patch creation. We created 512×512 non-overlapping patches from each original and corresponding ground truth image. Then, we applied horizontal flip, vertical flip, random flip, rotation, grid distortion, optical distortion, and elastic transforms on each image patch to augment the dataset further discussed in Sect. 3.2. We trained the chosen pretrained DeepLabV3 model on the augmented dataset and got the test result shown in Table 4.

Post-Challenge Submission: During challenge submission, the performance of the model was not desirable specially for segmenting the Intraretinal Microvascular abnormalities (IRMAs) shown in Table 4. In a 1024×1024 ground truth image of IMRAs, the minimum, maximum and average blob size of foreground pixels are 292, 68057 and 8612 respectively. Also, the foreground pixels are very scattered in nature. The comparative information about the foreground pixel blob size in ground truth image of IMRAS, NPAs and NV is given in Table 1. The creation of patches of size 512×512 lead to many empty patches. Training the model on this data gave poor generalization performance. Also, the features of NPAs in the original image corresponding to the ground truth are not very recognizable. So, for NPAs, the creation of the image patch made the feature representation capability of the model worse.

To handle these problems, we augmented the data by resizing it to 512×512 instead of creating a patch during post-challenge submission described in Subsect. 3.2. Resizing of image increased the model performance for all three types of lesions segmentation. Also, early stopping criteria during the training helped the model overcome some amount of overfitting.

Table 4. Performance of the model on the test dataset during challenge submission and post-challenge submission.

Submission Type	DSC1(%) (IMRAs)	DSC2(%) (NPAs)	DSC3(%) (NVs)	Mean DSC (%)	Mean IoU (%)
During Challenge Submission	06.73	45.18	44.27	32.06	22.05
Post Challenge Submission	25.40	60.18	44.51	43.36	31.03

6   Conclusion

This paper presented our methodology and results on DRAC22 datasets for both submissions. On the one hand, the dataset has a few numbers of data samples. On the other, a deep complex model needs a huge number of data to train. Therefore, we augmented the data and used transfer learning approach to segment Diabetic Retinopathy lesions from UW-OCTA images. Here, the application of different data augmentation techniques showed performance variation on the test set.

Although we have not applied any image enhancement techniques, but hope the application of the image enhancement technique may help to segment those lesions where lesion features are not representative. Also, proper customization of the pretrained model concerning the number of channels, depth of layers, dimension reduction techniques, etc., may help us achieve a good result in various medical image segmentation tasks in the future.

References

1. Idf diabetes atlas. https://www.idf.org/our-activities/care-prevention/eye-health.html. Accessed 05 Oct 2022
2. International diabetes federation. https://www.idf.org/our-activities/care-prevention/eye-health.html. Accessed 29 Sept 2022
3. Chen, L.C., Papandreou, G., Kokkinos, I., Murphy, K., Yuille, A.L.: Deeplab: semantic image segmentation with deep convolutional nets, atrous convolution, and fully connected crfs. IEEE Trans. Pattern Analy. Mach. Intell. **40**(4), 834–848 (2018). https://doi.org/10.1109/TPAMI.2017.2699184
4. Chen, L.C., Papandreou, G., Schroff, F., Adam, H.: Rethinking atrous convolution for semantic image segmentation. arXiv preprint arXiv:1706.05587 (2017)
5. Dai, L., et al.: A deep learning system for detecting diabetic retinopathy across the disease spectrum. Nat. Commun. **12**(1), 1–11 (2021)
6. Deng, J., Dong, W., Socher, R., Li, L.J., Li, K., Fei-Fei, L.: Imagenet: a large-scale hierarchical image database. In: 2009 IEEE Conference on Computer Vision and Pattern Recognition, pp. 248–255 (2009). https://doi.org/10.1109/CVPR.2009.5206848
7. He, K., Zhang, X., Ren, S., Sun, J.: Deep residual learning for image recognition. In: Proceedings of the IEEE Conference on Computer Vision and Pattern Recognition, pp. 770–778 (2016)
8. He, K., Zhang, X., Ren, S., Sun, J.: Deep residual learning for image recognition. In: The IEEE Conference on Computer Vision and Pattern Recognition (CVPR) (2016)
9. Huang, G., Liu, Z., Weinberger, K.Q.: Densely connected convolutional networks. CoRR abs/1608.06993 (2016). http://arxiv.org/abs/1608.06993
10. Liu, R., et al.: Deepdrid: diabetic retinopathy-grading and image quality estimation challenge. Patterns 100512 (2022)
11. Shelhamer, E., Long, J., Darrell, T.: Fully convolutional networks for semantic segmentation. IEEE Trans. Pattern Anal. Mach. Intell. **39**(4), 640–651 (2017). https://doi.org/10.1109/TPAMI.2016.2572683
12. Sheng, B., et al.: An overview of artificial intelligence in diabetic retinopathy and other ocular diseases. Front. Publ. Health **10** (2022)
13. Sheng, B., et al.: Diabetic retinopathy analysis challenge 2022 (2022). https://doi.org/10.5281/zenodo.6362349
14. Simonyan, K., Zisserman, A.: Very deep convolutional networks for large-scale image recognition. arXiv preprint arXiv:1409.1556 (2014)
15. Sultana, F., Sufian, A., Dutta, P.: Evolution of image segmentation using deep convolutional neural network: A survey. Knowl.-Based Syst. **201**, 106062 (2020). https://doi.org/10.1016/j.knosys.2020.106062. https://www.sciencedirect.com/science/article/pii/S0950705120303464
16. Szegedy, C., et al.: Going deeper with convolutions. In: The IEEE Conference on Computer Vision and Pattern Recognition (CVPR) (2015)

MIDOG

MIDOG 2022 Preface

The identification of mitotic figures in histological specimens is important for determination of tumor malignancy and is therefore a common step in histological assessments; however, this task is also notorious for having low reproducibility and high inter-rater discordance. With digital pathology pipelines being introduced in an increasing number of pathology laboratories, it becomes sensible to use machine learning algorithms to aid human observers. Automatic mitotic figure detection is a topic of ongoing high interest in computer vision research and computer-aided diagnostics over the last years, but is also strongly dependent on the visual representation of the tissue in the digital image. As shown in the first mitosis domain generalization challenge (MIDOG) in 2021, the digitization device (most often a whole-slide image scanner) is one key component contributing to a covariate shift in image representation. Machine learning algorithms that do not account for variance introduced by the shift of imaging devices will be likely to show dramatically decreased performance when applied on images from a different image acquisition device. In the MIDOG 2021 challenge, the participants were able to show that a combination of augmentation, domain-adversarial training and other techniques can improve domain robustness toward the scanner domain drastically. These findings are extremely relevant in order to enable application of algorithms in different laboratories that use different imaging devices.

Yet, histopathology is subject to more domain shifts than the imaging device. In the second version of the MIDOG challenge, held in conjunction with the 2022 International Conference on Medical Image Computing and Computer Assisted Intervention (MICCAI), we tackled the problem of robustness of mitosis detection against variations introduced by different tumor types and species, and thereby implicitly against variations introduced in different laboratories. Robustness of algorithms against these additional domain shifts needs to be achieved for a wide application in various diagnostic settings. The training set of this challenge included 403 2mm^2 region of interest-cropouts from digitized histopathological specimen from five different annotated tumor domains plus one unannotated tumor domain[1]. The training data was made available on April 20th, 2022. The evaluation was carried out on 100 images with ten disjoint tumor domains from various lab environments (tissue processing and imaging devices) and species (humans, dogs and cats). The challenge was organized by a group of scientists from Germany, the Netherlands and Austria.

One main target of the challenge was to investigate domain robustness for truly unseen domains with a highly variable image representation. The participants provided their algorithmic solutions on the basis of a docker submission on the grand-challenge.org platform. To test for the proper function of the container, the organizers made available a preliminary test set, including three previously unseen tumor domains and one seen tumor domain (human breast cancer) acquired on a different scanner. None of the tumor

---

[1] The data is publicy available on Zenodo via the following link:https://zenodo.org/record/654 7151#..

domains of this preliminary test set reoccurred in the challenge's test set. The organizers furthermore did not reveal any details about characteristics of the test set until after the challenge was over.

As last year, the organizers made available reference implementations of domain-adversarial approaches, provided by Frauke Wilm for the challenge. The approach was trained on the MIDOG 2021 and 2022 datasets and the respective results were made available to the participants at the beginning of the submission phase as an orientation. Additionally, the organizers made available the results of a state-of-the-art instance segmentation pipeline based on Mask-RCNN and trained on the MIDOG 2022 training dataset with standard augmentation techniques, which was developed by Jonas Ammeling and Jonathan Ganz for this purpose.

The participants were able to submit to two tracks: The first track restricted the use of data to the datasets provided by the challenge organizers. This also prohibited the use of any further knowledge to extend the dataset, such as by deriving further labels by a model trained on external data. The second track allowed for the use of any additional data in the training of the approaches, given that this data was already publicly available or was made publicly available to the participants one month prior to the final deadline of the challenge. This also incorporated the use of any models trained on additional data to generate further labels for the training data, e.g., the unlabeled tumor domain.

In total, 15 teams submitted to the preliminary phase of the challenge, which was open between August 5 and August 25, 2022. For the final submission phase, which was open between August 26 and August 29, 10 teams submitted their approach to track 1 (without additional data), and two teams submitted their approach to track 2 (additional data permitted). The participants were allowed one submission per day on the preliminary dataset and a single submission on the final test set. Along with this final submission, all participants had to hand in the link to a publicly available pre-print short paper describing their approach in sufficient detail. All of these short papers were subjected to single-blinded peer review by three reviewers, checking for methodological soundness and quality of the description.

All approaches submitted to the final test set passed the quality assessment and were invited to contribute to these proceedings. The three top-performing teams were invited to contribute an extended version of their paper. Finally, 6 teams decided to contribute a manuscript to the proceedings. These manuscripts were again subject to three independent reviews, in which for two papers a major revision was requested, two papers were accepted pending minor changes and two papers were accepted in the form they were handed in.

We would like to express our deep gratitude to the talented and hard-working people that made the MIDOG 2022 challenge possible. Most notably, these were the co-organizers of the challenge, in alphabetical order: Samir Jabari, Nikolas Stathonikos and Mitko Veta. We also are grateful to the pathologists who helped in the annotation process, Taryn Donovan and Robert Klopfleisch, as well as Frauke Wilm, Jonas Ammeling and Jonathan Ganz for implementing the baseline approaches. We thank grand-challenge.org for, again, providing their platform free of charge for our challenge. Further, we would like to thank Siemens Healthineers and Tribun Health, who donated the prizes for the top three participants in each category, and Data Donors e.V., who helped in the organization of the prize funding. We further received support from Schwarzman AMC New York to account for additional staining costs. This challenge would not have been possible without support by our institutions, which provided access to compute capabilities and the diagnostic archives to retrieve hundreds of tissue samples. Finally, we would like to thank all participants, without whom the second iteration of the MIDOG challenge would not have been such a wonderful continuation and great success.

Marc Aubreville
Katharina Breininger
Christof A. Bertram

MIDOG 2022 Organization

General Chair

Marc Aubreville   Technische Hochschule Ingolstadt,
                                    Germany

Organization Committee

Katharina Breininger   Friedrich-Alexander-Universität
                                   Erlangen-Nürnberg, Germany
Christof A. Bertram   University of Veterinary Medicine Vienna,
                                   Austria
Samir Jabari   Universitätsklinikum Erlangen,
                                   Friedrich-Alexander-Universität
                                   Erlangen-Nürnberg, Germany
Nikolas Stathonikos   University Medical Center Utrecht,
                                   The Netherlands
Mitko Veta   Technical University Eindhoven,
                                   The Netherlands

Data and Technical Contributors

Taryn A. Donovan   Schwarzman Animal Medical Center, USA
Frauke Wilm   Friedrich-Alexander-Universität
                                   Erlangen-Nürnberg, Germany
Jonas Ammeling   Technische Hochschule Ingolstadt,
                                   Germany
Jonathan Ganz   Technische Hochschule Ingolstadt,
                                   Germany
Robert Klopfleisch   Freie Universität Berlin, Germany

Additional Reviewers

Rutger Fick
Tribun Health, France

Reference Algorithms for the Mitosis Domain Generalization (MIDOG) 2022 Challenge

Jonas Ammeling[1]($\boxtimes$), Frauke Wilm[2,3], Jonathan Ganz[1], Katharina Breininger[3], and Marc Aubreville[1]

[1] Technische Hochschule Ingolstadt, Ingolstadt, Germany
jonas.ammeling@thi.de
[2] Pattern Recognition Lab, Computer Sciences, Friedrich-Alexander-Universität Erlangen-Nürnberg, Erlangen, Germany
[3] Department Artificial Intelligence in Biomedical Engineering, Friedrich-Alexander-Universität Erlangen-Nürnberg, Erlangen, Germany

Abstract. Robust mitosis detection on images from different tumor types, pathology labs, and species is a challenging task that was addressed in the MICCAI Mitosis Domain Generalization (MIDOG) 2022 challenge. In this work, we describe three reference algorithms that were provided as a baseline for the challenge: A Mask-RCNN-based instance segmentation model trained on the MIDOG 2022 dataset, and two different versions of the domain-adversarial RetinaNet which already served as the baseline for MIDOG 2021 challenge, one trained on the MIDOG 2022 dataset and the other trained only on human breast carcinoma from MIDOG 2021. The domain-adversarial RetinaNet trained on the MIDOG 2022 dataset had the highest F_1 score of 0.7135 on the final test set. When trained on breast carcinoma only, the same network had a much lower F_1 score of 0.4719, indicating a significant domain shift between mitotic figure and tissue representation in different tumor types.

1 Introduction

The mitotic count (MC) is a well-established biomarker for tumor proliferation and is typically assessed by counting mitotic figures in a selected field of interest. Methods for automatic quantification of the MC exist but their performance is sensitive to a change in the visual representation between training images and images encountered during inference [1]. The Mitosis Domain Generalization (MIDOG) 2021 challenge [2] was organized to address a covariate shift caused by different image acquisition devices, i.e., different whole slide scanners. The MIDOG 2022 challenge extended this by not only addressing different scanning systems, but also different tumor types and species which result in different cell morphologies and tissue architectures. We provided three reference algorithms for the MIDOG 2022 challenge that used different approaches to robustly assess the MC under this covariate shift. The performance of the reference algorithms was intended to serve as a suitable baseline for the participants of the challenge implementing state-of-the-art approaches.

© The Author(s), under exclusive license to Springer Nature Switzerland AG 2023
B. Sheng and M. Aubreville (Eds.): MIDOG 2022/DRAC 2022, LNCS 13597, pp. 201–205, 2023.
https://doi.org/10.1007/978-3-031-33658-4_19

2   Materials and Methods

2.1   Dataset

The MIDOG 2022 dataset is a multi-tumor, multi-species, multi-laboratory, and multi-scanner dataset. The training subset consists of a total of 404 regions of interest of size $2\,mm^2$ from digitized microscopy slides covering six tumor types: canine cutaneous mast cell tumor (CCMCT), canine lung carcinoma, canine lymphosarcoma, human melanoma (HMEL), human neuroendocrine tumor, and human breast carcinoma (HBC). The HBC cases were previously part of the MIDOG 2021 challenge dataset [2]. No annotations were provided for the HMEL cases to support the development of unsupervised domain adaption approaches.

A preliminary test set was provided for a sanity check of submitted algorithms prior to the final test set submission. The preliminary test set consists of five cases each for four different tumor types (canine osteosarcoma, canine pheochromocytoma, human lymphoma, and HBC). The tumor types in this subset were not part of the final test set to prevent optimization for the test set based on the preliminary test set.

The official test set consists of 100 cases, equally split between 10 different tumor types and acquired from four different pathology labs. Neither images nor tumor types were revealed to the challenge participants. The test set includes CCMCT, canine hemangiosarcoma, canine mammary carcinoma, human astrocytoma, human bladder carcinoma, human colon carcinoma, human melanoma, human meningioma, feline lymphosarcoma, and feline soft tissue sarcoma. The CCMCT cases of the training and test were from different labs and scanned using different scanners.

2.2   DA-RetinaNet

The DA-RetinaNet was proposed by Wilm *et al.* [9] and extends the approach of the RetinaNet object detector [7] with a domain-adversarial branch. The approach was previously used as reference algorithm for the MIDOG 2021 challenge, where it achieved an F_1 score of 0.7183 on the official test set, containing HBC cases from different scanners, placing the approach in the Top-5. The domain adversarial training was slightly modified for the MIDOG 2022 challenge. Instead of performing domain adversarial training over different scanner domains, the different domains were considered by different tumor types. Consequently, the loss function was calculated as the sum of the domain loss for all tumor types and the bounding box regression and instance classification loss for all cases except HMEL. In addition to the standard online augmentation pipeline, a stain augmentation technique based on Macenko's method for stain deconvolution [8] was used to generate a wider range of realistic stain variations during training. The network was trained with stochastic gradient descent (SGD) and a cyclic maximum learning rate of 0.0001 until convergence on the validation set. Model selection was performed by choosing the model with the highest performance on the validation set as well as the highest domain classification loss to ensure tumor type independent features within the model.

2.3   Mask-RCNN

Since the best-performing methods in MIDOG 2021 [3,5,10] used either segmentation or instance segmentation algorithms, we implemented a Mask-RCNN [4] for mitosis detection as a state-of-the-art instance segmentation model. We used the off-the-shelf implementation of Mask-RCNN from the Torchvision framework. We did not use specific domain adaptation or adversarial training techniques to provide participants with a baseline performance of what can be achieved in this way. The non-annotated HMEL cases were omitted from training. The network was trained with SGD until convergence on the validation set. The learning rate was increased linearly to 0.001 over the first two epochs and then reduced by a factor of 0.5 if the validation loss did not improve for more than five epochs. Since the training of Mask-RCNN requires pixel-level annotations, which were not provided as part of the challenge, we used the publicly available cell segmentation algorithm NuClick [6] to automatically obtain segmentation masks for all mitotic figures.

2.4   Network Training

In order to improve comparability between the two approaches we used the same training and (internal) validation split of the training dataset across tumor types and ensured a similar distribution of high and low MC cases in each subset. Both approaches were trained on patches of size 512×512 using standard online augmentations with the addition of stain augmentation for the DA-RetinaNet. Both approaches used a sampling strategy to overcome the high class imbalance of mitotic figures and similar-looking cells, with half of the patches sampled in a 512 px radius around a mitotic figure and the other half sampled either in a 512 px radius around a hard negative look-alike or completely randomly.

3   Evaluation

Training for both models was repeated three times and model selection was performed on all annotated tumor types in the validation subset. The final model for DA-RetinaNet and Mask-RCNN was selected based on the highest performance on the validation subset and the optimal detection threshold was set for each model accordingly. The final DA-RetinaNet achieved an F_1 score of 0.7441 on the internal validation subset and the final Mask-RCNN achieved an F_1 score of 0.7815. These two models were submitted as reference approaches to the MIDOG 2022 challenge using a Docker-based submission system [2]. Prior to evaluation on the final test set, both submissions were tested for proper functionality on the preliminary test set. Furthermore, we evaluated the DA-RetinaNet reference algorithm from the previous MIDOG 2021 challenge, which was trained solely on HBC and without stain augmentation on the MIDOG 2021 dataset, for comparison.

Table 1. Performance metrics on the MIDOG 2022 preliminary and final test set.

Approach	Preliminary test set		Final test set	
	mAP	F_1 score	mAP	F_1 score
DA-RetinaNet 2021 [9]	0.2995	0.4719	0.3916	0.5133
DA-RetinaNet 2022	**0.6102**	**0.7152**	**0.5518**	**0.7135**
Mask RCNN 2022	0.4933	0.6285	0.4708	0.6542

4 Results and Discussion

Table 1 summarizes the test set performance of the reference approaches. On the preliminary test set, DA-RetinaNet 2022 achieved an F_1 score of 0.7152 and Mask-RCNN an F_1 score of 0.6285. The 2021 challenge baseline (DA-RetinaNet 2021) achieved an F_1 score of 0.4719. On the final test set, DA-RetinaNet 2021 achieved the lowest results with an F_1 score of 0.5133, Mask-RCNN achieved an F_1 score of 0.6542, and DA-RetinaNet 2022 achieved the best performance with an F_1 score of 0.7153.

The results highlight the effectiveness of domain-adversarial training over the different tumor types. The lowest performance of the DA-RetinaNet 2021 indicates a strong domain shift between mitotic figures from images of different tumor types and tissue preparation procedures. Simply training on a larger variety of tumor types with the Mask-RCNN helped to improve the generalization performance on unseen cases but could not compete with the performance when using tailored domain-adversarial training as with DA-RetinaNet 2022.

References

1. Aubreville, M.: Quantifying the scanner-induced domain gap in mitosis detection. In: Proceedings of MIDL (2021)
2. Aubreville, M., et al.: Mitosis domain generalization in histopathology images - the MIDOG challenge. Med. Image Anal. **84**, 102699 (2023)
3. Fick, R.H.J., Moshayedi, A., Roy, G., Dedieu, J., Petit, S., Hadj, S.B.: Domain-specific cycle-GAN augmentation improves domain generalizability for mitosis detection. In: Aubreville, M., Zimmerer, D., Heinrich, M. (eds.) MICCAI 2021. LNCS, vol. 13166, pp. 40–47. Springer, Cham (2022). https://doi.org/10.1007/978-3-030-97281-3_5
4. He, K., et al.: Mask R-CNN. IEEE Trans. Pattern Anal. Mach. Intell. **42**, 386–397 (2017)
5. Jahanifar, M., et al.: Stain-robust mitotic figure detection for the mitosis domain generalization challenge. In: Aubreville, M., Zimmerer, D., Heinrich, M. (eds.) MICCAI 2021. LNCS, vol. 13166, pp. 48–52. Springer, Cham (2022). https://doi.org/10.1007/978-3-030-97281-3_6
6. Koohbanani, N.A., et al.: Nuclick: a deep learning framework for interactive segmentation of microscopic images. Med. Image Anal. **65**, 101771 (2020)
7. Lin, T.Y., et al.: Focal loss for dense object detection. IEEE Trans. Pattern Anal. Mach. Intell. **42**, 318–327 (2017)

8. Macenko, M., et al.: A method for normalizing histology slides for quantitative analysis. In: Proceedings of ISBI 2009, pp. 1107–1110. IEEE (2009)
9. Wilm, F., Marzahl, C., Breininger, K., Aubreville, M.: Domain adversarial RetinaNet as a reference algorithm for the MItosis DOmain generalization challenge. In: Aubreville, M., Zimmerer, D., Heinrich, M. (eds.) MICCAI 2021. LNCS, vol. 13166, pp. 5–13. Springer, Cham (2022). https://doi.org/10.1007/978-3-030-97281-3_1
10. Yang, S., Luo, F., Zhang, J., Wang, X.: Sk-Unet model with Fourier domain for mitosis detection. In: Aubreville, M., Zimmerer, D., Heinrich, M. (eds.) MICCAI 2021. LNCS, vol. 13166, pp. 86–90. Springer, Cham (2022). https://doi.org/10.1007/978-3-030-97281-3_14

Radial Prediction Domain Adaption Classifier for the MIDOG 2022 Challenge

Jonas Annuscheit$^{(\boxtimes)}$ and Christian Krumnow

University of Applied Sciences (HTW) Berlin, Center for biomedical image
and information processing, Ostendstraße 25, 12459 Berlin, Germany
Jonas.Annuscheit@HTW-Berlin.de

Abstract. This paper describes our contribution to the MIDOG 2022
challenge for detecting mitotic cells. One of the major problems to be
addressed in the MIDOG 2022 challenge is the robustness under the
natural variance that appears for real-life data in the histopathology
field. To address the problem, we use an adapted YOLOv5s model for
object detection in conjunction with a new Domain Adaption Classifier
(DAC) variant, the Radial-Prediction-DAC, to achieve robustness under
domain shifts. In addition, we increase the variability of the available
training data using stain augmentation in HED color space. Using the
suggested method, we obtain a test set F1-score of 0.6658.

Keywords: MIDOG 22 contribution · mitosis detection · Domain
Adaption Classifier · Stain Augmentation · Test Time Augmentation

1 Introduction

Deep Learning can help to reduce the time-consuming parts like counting mitoses
on H&E histology slides and can help to decrease the variability of the annotations [3]. In the used dataset of the MIDOG 2022 challenge [2], unseen tissue
types are available in the test dataset, leading to a significant domain shift from
the available training set to the test set.

In this work, we adapt a domain adaption classifier (DAC) by replacing
the softmax layer with a modified version of the Radial Prediction (RP) Layer
introduced in [4]. With our variant we are able to adaptively learn prototypes of
each scanner, tissue type, and case id. These prototypes represent the essential
collective features of each individual scanner, tissue or case within an abstract
embedding space and are then used to remove this information from the detection
model.

2 Material and Method

2.1 Dataset

The training dataset of the challenge [2] consists of 9501 point annotations of
mitotic figures from 403 $2mm^2$ regions of H&E stained tissue from four scanners
and five different tissue types, and one tissue type without annotations.

B. Sheng and M. Aubreville (Eds.): MIDOG 2022/DRAC 2022, LNCS 13597, pp. 206–210, 2023.
https://doi.org/10.1007/978-3-031-33658-4_20

The dataset was split balanced for each tissue type and scanner combination into 80% for the training set, 10% for the validation set, and 10% for an internal test set. For normalization, each $2mm^2$ region was resized to the highest resolution available in the dataset, and 30 overlapping patches of 1280×1280 pixels were created. To reduce the number of patches for each region, only one-third of the patches, with the most annotations of mitotic figures, were used for the training of the network. The unlabeled data was included in the training process to train the RP-DAC and to remove the domain information from the network.

2.2 Data Augmentation

Train Time Augmentation: To enlarge the source domain, we used the technique introduced by Tellez et al. [8] that scales and shifts the color channels in the HED color space to imitate color variations from different scanners. We use deconvolution [7] to convert an image from RGB to HED, multiply the hematoxylin, eosin, and dab channel, where the latter collects residual parts of the transformation, with three factors (α_H, α_E, α_D), and transform back to RGB. For each scanner/tissue combination, the values for each of the α's are drawn from a scaled and translated beta distribution that was manually selected. Distributions were chosen by first obtaining the interval spanned by the mean of 100 sampled images of each scanner/tissue combination in HED space in each channel. We then design for each channel a beta distribution that roughly resembles the distribution of the means of all images in the interval. The means of each scanner/tissue combination typically cover only a part of the interval in each channel. The beta distribution was then scaled for each scanner/tissue combination appropriately, such that the mean of the transformed images best resemble the distribution of the means of all images. By this, we enlarge the typical colorspace of a given scanner/tissue combination such that it can also resemble the color palette of a different scanner/tissue combination. Furthermore, we allow for randomly applied horizontal and vertical flips, reflection, up to full rotation, and translation up to 200 pixels. Furthermore, we randomly apply a weak blur or sharpening on the input image.

Test Time Augmentation: We mirrored the test images to use our trained YOLO model on four image variants. The corresponding four predictions are then combined as described in Sect. 2.3 in the evaluation part.

2.3 Model and Training

Detection Model: Our base model for the detection task was the YOLOv5s [5] model that was pretrained on the COCO dataset. The YOLOv5s model was used because it is a very fast model allowing for object detection with bounding boxes and has a strong community that keeps it updated with modern developments.

The YOLOv5s architecture contains multiple places where the output of two consecutive convolution layers is combined by concatenating the channels.

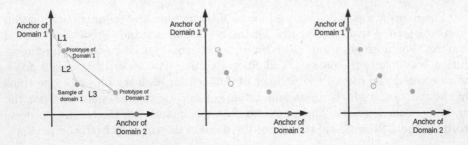

Fig. 1. Schematic visualization of the training in the n-dimensional space of the RP-DAC for $n = 2$. Left panel: initial configuration with two prototypes (green and blue) and one prediction (orange) of an input from domain 1. L_1, L_2 and L_3 denote the distance of the prototype and its anchor or the prediction as indicated. Center panel: during the RP-DAC update (step 1 of the training) distance of prototype and prediction is reduced by minimizing $0.1L_1^2 + 0.9L_2^2$). Right panel: during a YOLO-update step (step 2 of the training) the weights of the YOLO model are adapted such that the prediction tends towards the center of the prototypes (minimizing the sum $L_2^2 + L_3^2$ (Color figure online)

At these places, we introduce individually learnable parameters p that allow reducing the influence of the earlier of the two convolutional layers by scaling its output with sigmoid(p) for the concatenation. A detailed list of the used layers and the implementation of the model can be found at [1].

RP-DAC: In [4] the authors developed the Radial Prediction Layer (RPL). For a classification problem with n classes, the RPL is used as a replacement for the softmax layers. It uses n fixed points in an n-dimensional space as prototypes (one for each class) and measures the distance of each (embedded) input to each prototype - allowing them to minimize the distance to the prototype of the corresponding class during training. We modified the idea of the RPL such that it can be used as a DAC with moving prototypes P_i that are linked to fixed anchor A_i. Considering n domain classes, we adapt the n-dimensional RPL by using the n unit-vectors $\hat{e}_i$ as the fixed anchors A_i and initialize the moving prototypes P_i on the corresponding n with 0.1 scaled unit-vectors, i.e., $P_i = 0.1\,\hat{e}_i$. Each prototype P_i is then linked to the anchor A_i in the training process.

The DAC model itself gets the same input as the detection head of the YOLOv5s model, where we upscale all three feature maps to the largest size occurring and concatenate them along the channels, The DAC then consists of a 1×1 convolution layer that reduces the number of channels to 64, two successive 3×3 convolution layers with stride 2, and a global average pooling. The resulting feature vector is copied and passed to three linear layers that map it into a $n = 4$, $n = 6$ and $n = 403$ dimensional space, respectively ($n = 4$ when each scanners, $n = 6$ when each tissue types and $n = 403$ when each case id constitutes a different domain), followed each by our adapted RP-layer with moving prototype with the corresponding dimension. The DAC is trained (in conjunction with the YOLO model as explained below) to classify the different classes in the domain (e.g. identify the scanner of a given image). During that training, the prototypes can move freely in the n-dimensional space while the deviation from the corresponding anchor point enters into the loss.

Training: The training of the whole model is divided into two alternating steps using a batch of $N = 64$ accumulated patches (8 mini-batches of size 8) and an AdamW optimizer [6] for 800 epochs with learning rates starting at 0.002 and reducing the learning rate by using an OneCycleLR. **1.** A training step for the RP-DACs for a given batch B of inputs x with labels l_1, l_2 and l_3 (where l_1 denotes the scanner, l_2 the tissue type and l_3 the case id of x) uses the total loss

$$\mathcal{L}_{\mathrm{RP-DAC}} = \sum_{i=1,2,3} \frac{1}{N\,n_i} \sum_{(x,l_i)\in B} \left[0.1 \cdot \|p_l^{(i)} - a_l^{(i)}\|_2^2 + 0.9 \cdot \|p_l^{(i)} - z^{(i)}\|_2^2 \right]$$

with n_i being the dimension, $p_l^{(i)}$ and $a_l^{(i)}$ are the prototype and anchor corresponding to the label l and $z^{(i)}$ denotes the prediction for the not augmented input x of the i-th RP-DAC. Here, only the weights of the RP-DAC layers and the position of the prototypes are updated. **2.** A training step for the YOLOv5 model uses the described train-time augmentation and bounding boxes of fixed size around each annotation. The total loss of this training step is given by

$$\mathcal{L}_{\mathrm{detection}} = \mathcal{L}_{\mathrm{YOLOv5}} + \sum_{i=1,2,3} \frac{1}{N\,n_i} \sum_{l=1}^{n_i} \|p_l^{(i)} - z^{(i)}\|_2^2$$

where $\mathcal{L}_{\mathrm{YOLOv5}}$ denotes the standard YOLOv5 loss and the second term is the averaged distance of the prediction $z^{(i)}$ for x of the i-th RP-DAC to all prototypes. Here, only the weights of the YOLOv5 model are updated. The RP-DAC contribution leads the YOLOv5 model to produce representations of the inputs such that their RP-DAC predictions lies in the center of all prototypes, i.e. the YOLOv5 representation is maximally agnostic to the domain (see Fig. 1).

Evaluation: During the evaluation, we use the described test time augmentation and obtain for each input image four sets of predictions for bounding boxes of the detected mitosis and corresponding detection probabilities. The four predictions are combined into one set of point annotations by computing the average center of close by bounding boxes. A detected signal is then counted as mitosis if the detection probability is above the threshold of 0.403, which resulted in the best validation set F1-score.

3   Results

Our method yields an F1-score of 0.77, 0.74 and 0.75 on our training, validation and internal test set. On the not publicly available challenge test set, which contains multiple unseen tissue types we achieve an F1-score of 0.67. In order to gauge our method further, we performed an additional training in which the data from the Aperio ScanScope CS2 was left out completely (by this we also left out any patches containing canine cutaneous mast cell tumor in the training). On the data of the left out scanner we then achieve an F1-score of 0.74 while obtaining the F1-scores of 0.81, 0.74 and 0.75 on the corresponding reduced training, validation and test set. From the results of our experiments and on the challenge test set we conclude, that our scheme shows a promising robustness under domain shifts, the limits of which need to investigated in future research.

4   Discussion

In this work, we presented our contribution to the MIDOG challenge 2022. Most notably, we introduce the RP-DAC that utilizes moving prototypes instead of fixed classes for a DAC providing additional flexibility to account for instance for overlapping domains. We show that the RP-DAC can be trained stably in conjunction with a YOLOv5s detection model leading to promising domain transferability properties. Learning with such adaptable prototypes provides much potential for future work. Combining prototypes of similar classes using either merging strategies or other clustering methods or using additional domain adaption techniques, e.g., regarding the domain shift removal on residual connections [9] are promising routes.

Acknowledgements. J.A. acknowledges the financial support by the Federal Ministry of Education and Research of Germany project deep.Health (project number 13FH770IX6).

References

1. Github repository. https://github.com/JonasAnnuscheit/RPDAC_FOR_MIDOG22
2. Aubreville, M., Bertram, C., Breininger, K., Jabari, S., Stathonikos, N., Veta, M.: MItosis domain generalization challenge 2022. In: 25th International Conference on Medical Image Computing and Computer Assisted Intervention (MICCAI 2022) (2022). https://doi.org/10.5281/zenodo.6362337
3. Aubreville, M., et al.: Mitosis domain generalization in histopathology images - the MIDOG challenge. Med. Image Anal. **84**, 102699 (2023). https://doi.org/10.1016/j.media.2022.102699
4. Herta, C., Voigt, B.: Radial prediction layer. arXiv preprint: arxiv:1905.11150 (2019). https://doi.org/10.48550/arXiv.1905.11150
5. Jocher, G.: ultralytics/yolov5: v3.1 - Bug Fixes and Performance Improvements, October 2020. https://github.com/ultralytics/yolov5. https://doi.org/10.5281/zenodo.4154370
6. Loshchilov, I., Hutter, F.: Decoupled weight decay regularization. arxiv preprint: arXiv:1711.05101 (2017). https://doi.org/10.48550/arXiv.1711.05101
7. Ruifrok, A.C., Johnston, D.A.: Quantification of histochemical staining by color deconvolution. Anal. Quant. Cytol. Histol. **23**(4), 291–299 (2001)
8. Tellez, D., et al.: Whole-slide mitosis detection in H&E breast histology using PHH3 as a reference to train distilled stain-invariant convolutional networks. IEEE Trans. Med. Imag. **37**(9), 2126–2136 (2018)
9. Zheng, J., Wu, W., Zhao, Y., Fu, H.: Transresnet: transferable resnet for domain adaptation. In: 2021 IEEE International Conference on Image Processing (ICIP), pp. 764–768 (2021). https://doi.org/10.1109/ICIP42928.2021.9506562

Detecting Mitoses with a Convolutional Neural Network for MIDOG 2022 Challenge

Hongyan Gu[1], Mohammad Haeri[2], Shuo Ni[1],
Christopher Kazu Williams[3], Neda Zarrin-Khameh[4], Shino Magaki[3],
and Xiang 'Anthony' Chen[1]($\boxtimes$)

[1] University of California, Los Angeles, Los Angeles, USA
xac@ucla.edu
[2] University of Kansas Medical Center, Kansas City, USA
[3] UCLA David Geffen School of Medicine, Los Angeles, USA
[4] Baylor College of Medicine, Houston, USA

Abstract. This work presents a mitosis detection method with only one vanilla Convolutional Neural Network (CNN). Our method consists of two steps: given an image, we first apply a CNN using a sliding window technique to extract patches that have mitoses; we then calculate each extracted patch's class activation map to obtain the mitosis's precise location. To increase the model performance on high-domain-variance pathology images, we train the CNN with a data augmentation pipeline, a noise-tolerant loss that copes with unlabeled images, and a multi-rounded active learning strategy. In the MIDOG 2022 challenge, our approach, with an EfficientNet-b3 CNN model, achieved an overall F1 score of 0.7323 in the preliminary test phase, and 0.6847 in the final test phase (task 1). Our approach sheds light on the broader applicability of class activation maps for object detections in pathology images.

Keywords: Mitosis detection · Domain shift · Convolutional neural network · Class activation map

1 Introduction

Mitotic activity is a crucial histopathological indicator related to cancer malignancy and patients' prognosis [7]. Because of its importance, a considerable amount of literature has proposed datasets [2,3] and deep learning models [11,12] for mitosis detection. To date, a number of state-of-the-art deep learning approaches detect mitoses with the segmentation task [10,16]. However, these methods usually require generating pixel-level segmentation maps as ground truth. Another school of techniques utilizes object detection with a two-stage setup — a localization model (*e.g.*, RetinaNet) is employed for extracting interest locations, followed by a classification model to tell whether these locations

B. Sheng and M. Aubreville (Eds.): MIDOG 2022/DRAC 2022, LNCS 13597, pp. 211–216, 2023.
https://doi.org/10.1007/978-3-031-33658-4_21

have mitoses [2,11,12]. Such a two-stage setup was reported to improve the performance of mitosis detection compared to that with the localization model only [2].

Since adding a classification model can improve the performance, we argue that using only one CNN model for mitosis detection is also viable. Because CNNs cannot directly report the location of mitosis, previous works either modified the structure of CNNs [6], or used CNNs with a small input size to reduce the localization errors [14]. Instead, our approach extracts the location of mitoses with the class activation map (CAM) [17], which allows CNNs to accept a larger input size for more efficient training. Also, our approach is model-agnostic, and can work with vanilla CNNs because calculating CAMs does not require changing the network structure.

We validated our proposed method in MItosis DOmain Generalization (MIDOG) 2022 Challenge [1]. The challenge training set consists of 403 Hematoxylin & Eosin (H&E) regions of interest (ROIs, average size=5143×6860 pixels), covering six tumor types scanned from multiple scanners. 354/403 ROIs have been labeled and have 9,501 mitotic figures. The preliminary test set includes 20 cases from four tumor types, and the final test set has 100 independent tumor cases from ten tumor types. Given the dataset's high variance, we employed three techniques to improve the CNN's performance:

1. An augmentation pipeline with balanced-mixup [8] and stain augmentation [14];
2. The utilization of unlabeled images for training (treated as negative), and training with the Online Uncertainty Sample Mining (OUSM) [15] to gain robustness with noisy labels;
3. A multi-rounded, active learning training strategy that adds false-positive, false-negative, and hard-negative patches after each round of training.

2   Methods

2.1   Extracting Patches for Initial Training

We randomly used $\sim 90\%$ of the image instances in the MIDOG 2022 Challenge to generate the training set and $\sim 10\%$ for the validation set. To maximally utilize the dataset, we included unlabelled images in the training set and treated them as negative images (*i.e.,* no mitoses inside). For each image, we extracted patches with the size of $240 \times 240 \times 3$ pixels surrounding the location of each annotation (provided by the challenge) and placed them into the train/validation set.

2.2   Model Training

We trained an EfficientNet-b3 [13] model (input size: $240 \times 240 \times 3$) with pretrained ImageNet weights. Here, we tried to improve model performance by constructing an online data augmentation pipeline. The pipeline includes general

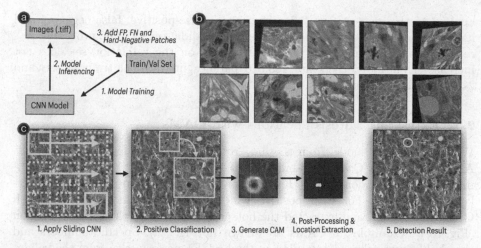

Fig. 1. (a) Illustration of the active learning training strategy used in this work; (b) Examples of augmented patches according to our augmentation pipeline; (c) Overall data processing pipeline our approach: detecting mitosis using a convolution neural network and the class activation map.

image augmentation techniques, including random rotation, flip, elastic transform, grid distortion, affine, color jitter, Gaussian blur, and Gaussian noise. Besides, we added two augmentation methods – stain augmentation [14] and balance-mixup [8] – to deal with the domain shift in pathology images. Examples of augmented patches are shown in Fig. 1(b). The model was trained with an SGD optimizer with momentum 0.9, a Cosine Annealing learning rate scheduler with warm restart (max LR=6×10^{-4}). Since we treated all unlabeled images as negative, we further used an OUSM [15] + COnsistent RAnk Logits (CORAL) loss [4] to deal with noisy labels. Each round of training had 100 epochs, and we selected the model with the highest F1 score on the validation set for inferencing.

2.3 Inferencing

We slid the trained EfficientNet on train and validation images with window size 240×240 and step-size 30. We then cross-referenced the CNN predictions with the ground truth. Here, we define a positive window classification as a true-positive if mitoses were inside the window and a false-positive otherwise. We further define false-negative if no positive windows surround a mitosis annotation.

2.4 Incrementing the Patch Dataset with Active Learning

We employed a multi-round active learning process to boost the performance of the EfficientNet (Fig. 1(a)). Each round starts with the model training on the current train/validation set (Sect. 2.2). Then, the best model is selected and applied to the images (Sect. 2.3). After that, false-positive, false-negative, and hard-negative patches are added to the train/validation set. The procedure was repeated six times until the model's F1 score on the validation set does not increase. Eventually, there are 103,816 patches in the final training set, and 23,638 in the validation set.

2.5   Extracting Mitosis Locations with CAMs

We used the best model from the final round in Sect. 2.4 for the test images. A window with a CNN probability > 0.84 was considered positive, and non-maximum suppression with a threshold of 0.22 was used to mitigate the overlapping windows. For each positive window, we calculated the CAM with GradCAM++ [5], and extracted the hotspot's centroid as the mitosis location (Fig. 1(c)). Examples of CAMs are shown in Fig. 2(a). Specific numbers and thresholds were selected according to the best F1 performance from the validation images.

3   Results

On the preliminary test phase of the MIDOG 2022 Challenge, our approach achieved an overall F1 score of 0.7323, with 0.7313 precision and 0.7333 recall. In the final test phase of the challenge (task 1), our approach achieved the F1 score of 0.6847 (precision:0.7559, recall: 0.6258). In sum, our approach is 2.34% higher than the baseline RetinaNet approach regarding the overall F1 score in the preliminary test phase, but 4.21% lower in the final test phase. Please refer to the Grand-Challenge Leader-board[1] for more details of the test result.

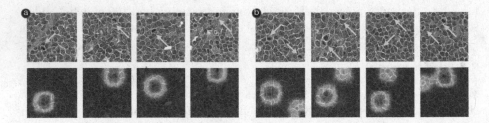

Fig. 2. Examples of positive patches detected by CNN and corresponding CAMs generated by GradCAM++ from image 240.tiff. (a) Patch samples and CAMs if there is one mitosis inside (pointed by the green arrow); (b) Patch samples and CAMs if there are multiple mitoses inside. Note that generated CAMs might not report strong signals for some mitoses.

---

[1] https://midog2022.grand-challenge.org/evaluation/final-test-phase-task-1-without-additional-data/leaderboard/.

4 Discussion and Conclusion

Although CAMs are primarily used for explaining CNN classifications, we demonstrate their potential usage to help detect mitoses in H&E images. We believe our approach has the potential to work with mobile/edge devices, which have limited computational power and are not optimized for special-structured deep-learning models.

It is noteworthy that CAMs might fail to highlight all mitoses when there are multiple in an image. As shown in Fig. 2(b), if there were multiple mitoses inside one positive patch, CAMs might report a strong signal for one mitosis, while giving weak signals for others. However, we argue that our approach *as-is* can still be helpful in diseases where mitosis is not a high-prevalent histological pattern (*e.g.,* meningiomas [9]).

On the other hand, we believe that the limitation is partly caused by using classification models for localization tasks: given a patch with multiple mitoses, a CNN only needs to find at least one to predict it as positive. As such, other mitoses are more or less ignored. To compensate for the limitation, we suggest future works to improve the quality of CAMs by aligning them with ground-truth mitosis location maps. For example, a "CAM-loss" might be designed to penalize the misalignment between the CAM and mitosis location heatmaps [18]. Different from the segmentation task, generating mitosis location maps does not require pixel-level segmentation masks: simply applying a Gaussian kernel over the locations of mitoses can generate a mitosis location map similar to Fig. 2.

Acknowledgement. This work was funded in part by the Office of Naval Research under grant N000142212188.

References

1. Aubreville, M., Bertram, C., Breininger, K., Jabari, S., Stathonikos, N., Veta, M.: Mitosis domain generalization challenge 2022. In: 25th International Conference on Medical Image Computing and Computer Assisted Intervention (MICCAI 2022) (2022). https://doi.org/10.5281/zenodo.6362337

2. Aubreville, M., Bertram, C.A., Donovan, T.A., Marzahl, C., Maier, A., Klopfleisch, R.: A completely annotated whole slide image dataset of canine breast cancer to aid human breast cancer research. Sci. Data **7**(1), 1–10 (2020)

3. Bertram, C.A., Aubreville, M., Marzahl, C., Maier, A., Klopfleisch, R.: A large-scale dataset for mitotic figure assessment on whole slide images of canine cutaneous mast cell tumor. Sci. Data **6**(1), 1–9 (2019)

4. Cao, W., Mirjalili, V., Raschka, S.: Rank consistent ordinal regression for neural networks with application to age estimation. Pattern Recogn. Lett. **140**, 325–331 (2020)

5. Chattopadhay, A., Sarkar, A., Howlader, P., Balasubramanian, V.N.: Grad-cam++: generalized gradient-based visual explanations for deep convolutional networks. In: 2018 IEEE Winter Conference on Applications of Computer Vision (WACV), pp. 839–847 (2018). https://doi.org/10.1109/WACV.2018.00097

6. Cireşan, D.C., Giusti, A., Gambardella, L.M., Schmidhuber, J.: Mitosis detection in breast cancer histology images with deep neural networks. In: Mori, K., Sakuma, I., Sato, Y., Barillot, C., Navab, N. (eds.) MICCAI 2013. LNCS, vol. 8150, pp. 411–418. Springer, Heidelberg (2013). https://doi.org/10.1007/978-3-642-40763-5_51

7. Cree, I.A., et al.: Counting mitoses: SI (ze) matters! Mod. Pathol. 34(9), 1651–1657 (2021)

8. Galdran, A., Carneiro, G., González Ballester, M.A.: Balanced-MixUp for highly imbalanced medical image classification. In: de Bruijne, M., et al. (eds.) MICCAI 2021. LNCS, vol. 12905, pp. 323–333. Springer, Cham (2021). https://doi.org/10.1007/978-3-030-87240-3_31

9. Gu, H., et al.: Improving workflow integration with xPath: design and evaluation of a human-AI diagnosis system in pathology (2021)

10. Jahanifar, M., et al.: Stain-robust mitotic figure detection for the mitosis domain generalization challenge. In: Aubreville, M., Zimmerer, D., Heinrich, M. (eds.) MICCAI 2021. LNCS, vol. 13166, pp. 48–52. Springer, Cham (2022). https://doi.org/10.1007/978-3-030-97281-3_6

11. Li, C., Wang, X., Liu, W., Latecki, L.J.: Deepmitosis: mitosis detection via deep detection, verification and segmentation networks. Med. Image Anal. 45, 121–133 (2018)

12. Mahmood, T., Arsalan, M., Owais, M., Lee, M.B., Park, K.R.: Artificial intelligence-based mitosis detection in breast cancer histopathology images using faster R-CNN and deep CNNs. J. Clin. Med. 9(3), 749 (2020)

13. Tan, M., Le, Q.: EfficientNet: rethinking model scaling for convolutional neural networks. In: International Conference on Machine Learning, pp. 6105–6114. PMLR (2019)

14. Tellez, D., et al.: Whole-slide mitosis detection in H&E breast histology using phh3 as a reference to train distilled stain-invariant convolutional networks. IEEE Trans. Med. Imaging 37(9), 2126–2136 (2018). https://doi.org/10.1109/TMI.2018.2820199

15. Xue, C., Dou, Q., Shi, X., Chen, H., Heng, P.A.: Robust learning at noisy labeled medical images: applied to skin lesion classification. In: 2019 IEEE 16th International Symposium on Biomedical Imaging (ISBI 2019), pp. 1280–1283. IEEE (2019)

16. Yang, S., Luo, F., Zhang, J., Wang, X.: Sk-Unet model with Fourier domain for mitosis detection. In: Aubreville, M., Zimmerer, D., Heinrich, M. (eds.) MICCAI 2021. LNCS, vol. 13166, pp. 86–90. Springer, Cham (2022). https://doi.org/10.1007/978-3-030-97281-3_14

17. Zhou, B., Khosla, A., Lapedriza, A., Oliva, A., Torralba, A.: Learning deep features for discriminative localization. In: Proceedings of the IEEE Conference on Computer Vision and Pattern Recognition, pp. 2921–2929 (2016)

18. Zhu, H., Salcudean, S., Rohling, R.: Gaze-guided class activation mapping: leveraging human attention for network attention in chest X-rays classification. arXiv preprint: arXiv:2202.07107 (2022)

Tackling Mitosis Domain Generalization in Histopathology Images with Color Normalization

Satoshi Kondo[1]($\boxtimes$) (iD), Satoshi Kasai[2], and Kousuke Hirasawa[3]

[1] Muroran Institute of Technology, Muroran, Hokkaido 050-8585, Japan
kondo@mmm.muroran-it.ac.jp
[2] Niigata University of Healthcare and Welfare, Niigata 950-3198, Japan
[3] Konica Minolta, Inc., Takatsuki, Osaka 569-8503, Japan

Abstract. In this paper, we propose a method for mitosis detection in histopathology images in an unsupervised domain adaptation setting. Our method is a two-step approach. The first step is color normalization, which is an unsupervised domain adaptation at the input level. In the second step, we use an object detection method for mitosis detection. Using a final test set consisting of patches from whole slide images and containing 100 independent tumor cases across 10 tumor types, we evaluate our method and obtain an F1 score of 0.671.

Keywords: Mitosis Detection · Domain Adaptation · Color normalization

1 Introduction

Mitosis detection is a key component of tumor prognosis for various tumors. In recent years, deep learning-based methods have demonstrated high human expert level accuracy in mitosis detection [2]. Mitosis is known to be relevant for many tumor types, but when a model is trained on a specific tumor or tissue type, model performance typically drops significantly on other tumor types.

The Mitosis Domain Generalization Challenge 2021 (MIDOG2021) [1] was conducted to evaluate the performance of different mitosis detection methods using breast cancer cases digitized by different scanners. MIDOG2022 was conducted as an extension of MIDOG2021. In MIDOG2022, histopathology images acquired from different laboratories, from different species (human, dog, cat) in the study area, and with five different whole slide scanners are newly provided. The training set consists of a total of 403 tumor cases from six tumor types, including canine lung cancer, human breast cancer, canine lymphoma, human neuroendocrine tumor, canine cutaneous mast cell tumor, and human melanoma. The test set contains 100 independent tumor cases across 10 tumor types.

In this paper, we propose a method for mitosis detection in histopathology images as a solution to MIDOG2022. We propose a kind of unsupervised

B. Sheng and M. Aubreville (Eds.): MIDOG 2022/DRAC 2022, LNCS 13597, pp. 217–220, 2023.
https://doi.org/10.1007/978-3-031-33658-4_22

domain adaptation (UDA) method. UDA is a type of domain adaptation and exploits labeled data from the source domain and unlabeled data from the target domain [4]. UDA approaches can be grouped based on where the domain adaptation is performed, at the input level, at the feature level, at the output level, or at the network levels, e.g., at the end of the feature extraction network, at intermediate levels, and so on [7]. Our proposed method performs the domain adaptation at the input level because the domain adaptation at the input level is simple and effective.

2   Proposed Method

Figure 1 shows an overview of our proposed method. Our method is a two-step approach. In the first step, we use color normalization as the input level domain adaptation. The color normalized image is divided into patches, and object detection is performed for each patch. The detection results for a given patch are combined to obtain the final detection results.

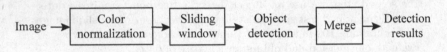

Fig. 1. Overview of our proposed method.

We use the method proposed in [8] for color normalization. First, the input image is decomposed into stain density maps that are sparse and non-negative. This is based on the assumption that tissue samples are stained with only a few stains, and most tissue regions are characterized by at most one effective stain. For a given image, its stain density maps are combined with the stain color base of the target image, thus changing only its color while preserving the structure of the described map.

The input histopathology images are very large, e.g. about $7,000 \times 5,000$ pixels. Therefore, after color normalization, the images are divided into patches because it is difficult to process the entire slide image at once. For each patch, we use an object detector to perform mitosis detection. We use EfficientDet [6] as our object detector, and the backbone of our EfficientDet is EfficientNetV2-L [5]. The mitosis detection results for patches are merged into an original space.

3   Experiments

The dataset used in our experiments includes digitized slides from 520 tumor cases, acquired from different laboratories, from different species (human, dog, cat), and with five different whole slide scanners. The training set consists of 400 cases and contains at least 50 cases each of prostate carcinoma, lymphoma,

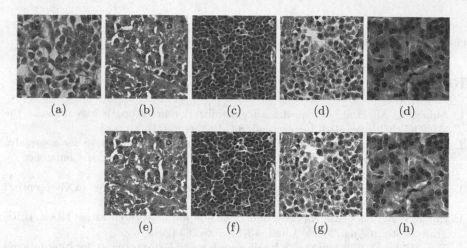

Fig. 2. Examples of color normalization results. (a) Template image for color normalization. (b)-(d) Original images. (e)-(h) Color normalized images. Each image in the bottom row corresponds to an image in the top row and the same column.

lung carcinoma, melanoma, breast cancer, and mast cell tumor. For each of the four different tumor types, the validation dataset contains five cases.

We manually select one image from the training dataset as a template for color normalization. Example color normalization results are shown in Fig. 2. The patch size is 512×512 pixels, and adjacent patches have region overlap in 67 pixels. Two types of objects are annotated in the dataset. One type is the "mitotic" class, and the other type is the "non-mitotic (hard negative)" class. Our object detector is trained to detect these two types of classes. In the inference, we use only the mitotic class and select the objects that have scores above a threshold of 0.5. When training the object detector, the learning rate is 0.002, the batch size is 8, and the number of epochs is 30. The optimizer is Adam [3]. We use horizontal and vertical flips as augmentation. However, we do not use unlabeled images as training data. Besides the dataset provided by the organizer of MIDOG2022, we do not use any additional data.

The evaluation is performed using a final test set. The final test set uses 10 tumor types with 10 cases each (100 cases in total). The experimental results show that the mAP is 0.466, the recall is 0.578, the precision is 0.798 and the F1 score is 0.671.

4   Conclusion

In this paper, we propose a method for mitosis detection in histopathology images in unsupervised domain adaptation setting. Our method is a two-step approach including color normalization, which is unsupervised domain adaptation at the input level, and object detection. We evaluated our method using the final test set containing 100 cases and we obtained an F1 score of 0.671.

Our future work is to evaluate how the performance of our proposed method changes depending on the choice of template for color normalization.

References

1. Aubreville, M., et al.: Mitosis domain generalization in histopathology images – the MIDOG challenge. Med. Image Anal. **84**, 102699 (2023)
2. Hussain, H., Hujran, O., Nitha, K.P.: Survey on mitosis detection for aggressive breast cancer from histological images. In: 2019 5th International Conference on Information Management (ICIM), pp. 232–236 (2019)
3. Kingma, D.P., Ba, J.: Adam: a method for stochastic optimization. arXiv preprint: arXiv:1412.6980 (2014)
4. Kouw, W.M., Loog, M.: A review of domain adaptation without target labels. IEEE Trans. Pattern Anal. Mach. Intell. **43**(3), 766–785 (2019)
5. Tan, M., Le, Q.: EfficientNetv2: smaller models and faster training. In: International Conference on Machine Learning, pp. 10096–10106. PMLR (2021)
6. Tan, M., Pang, R., Le, Q.V.: EfficientDet: scalable and efficient object detection. In: Proceedings of the IEEE/CVF Conference on Computer Vision and Pattern Recognition, pp. 10781–10790 (2020)
7. Toldo, M., Maracani, A., Michieli, U., Zanuttigh, P.: Unsupervised domain adaptation in semantic segmentation: a review. Technologies **8**(2), 35 (2020)
8. Vahadane, A., et al.: Structure-preserving color normalization and sparse stain separation for histological images. IEEE Trans. Med. Imag. **35**(8), 1962–1971 (2016)

A Deep Learning Based Ensemble Model for Generalized Mitosis Detection in H&E Stained Whole Slide Images

Sujatha Kotte[1], VG Saipradeep[1(✉)], Naveen Sivadasan[1], Thomas Joseph[1],
Hrishikesh Sharma[1], Vidushi Walia[1], Binuja Varma[2],
and Geetashree Mukherjee[3]

[1] TCS Research, Tata Consultancy Services Ltd., Hyderabad, India
{kotte.sujatha,saipradeep.v,naveen.sivadasan,thomas.joseph,
hrishikesh.sharma,vidushi.walia}@tcs.com
[2] Tata Consultancy Services Ltd, New Delhi, India
binuja.varma@tcs.com
[3] Tata Medical Center, Kolkata, India
geetashree.mukherjee@tmckolkata.com

Abstract. Identification of mitotic cells as well as estimation of mitotic index are important parameters in understanding the pathology of cancer, predicting response to chemotherapy and overall survival. This is usually performed manually by pathologists and there can be considerable variability in their assessments. The use of deep learning(DL) models can help in addressing this issue. However, most of the state-of-the-art methods are trained for specific cancer types, and often tend to fail when used across multiple tumor types. Hence there is a clear need for a more 'pan-tumor' approach to identifying mitotic figures. We propose a generalized DL model for mitosis detection using the MIDOG-2022 Challenge dataset. Using an ensemble of predictions from a transformer-based object detector and a separate classifier, our model makes final predictions. Our approach achieved an F1-score of 0.7569 and stood second in the MIDOG-2022 challenge. The predictions from the object detector alone achieved an F1-score of 0.7510. Our model generalizes well to address the domain shifts caused by variability in image acquisition, protocols and tumor tissue types.

Keywords: mitosis · digital pathology · H&E · vision transformers

1 Introduction

The most common method of identification of mitotic cells is manual examination of Hematoxylin and Eosin (H&E) stained tissue slides. With advances in the field of digital pathology [1], there have been developments in building highly

---

S. Kotte and V.G. Saipradeep—Joint first author.

B. Sheng and M. Aubreville (Eds.): MIDOG 2022/DRAC 2022, LNCS 13597, pp. 221–225, 2023.
https://doi.org/10.1007/978-3-031-33658-4_23

advanced scanners to capture images from H&E-stained whole slides. However, many state-of-the-art DL models trained on these whole slide images (WSIs) on a restricted set of tumor tissue types. Hence, their performance suffers due to domain shifts caused by unseen tumor tissue types. The inter-lab differences in staining protocols as well as differences in scanner outputs also affect the performance of many digital pathology methods. This has limited the adoption of computer-based tools by clinical pathologists [2].

To address these issues, there have been many challenges organized in the past, especially for mitosis detection, namely, TUPAC16 [3], Mitos & Atypia 14 [4], and MIDOG-2021 [5]. In the MIDOG 2021 challenge, the aim was to look at methods to help overcome the domain shift caused by variability in image acquisition. The MIDOG 2022 Challenge [6] looks to further develop and analyze the performance of approaches that help resolve the additional problem of domain-shift due to variability in tissue types. In this work, we propose a deep learning-based ensemble model that detects mitotic figures with improved generalization performance.

2 Methods

Fig. 1 depicts the overall architecture of our approach.

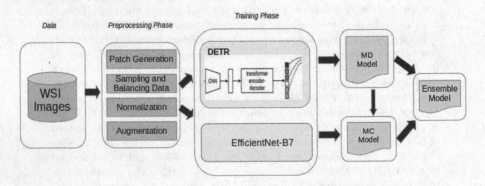

Fig. 1. Architecture of Ensemble Model

2.1 Data

The MIDOG 2022 organizers have provided non-overlapping training, preliminary and test sets of regions of interest (ROI) from WSIs prepared from different tumor cases. Details of the dataset are available at [7]. No other external datasets were used in our approach.

2.2 Data Pre-processing

Based on the 354-labeled images in the training dataset, separate training and validation datasets were created using 80:20 split. We observed that the MIDOG

2022 dataset has more of mitotic look-alikes (imposters) than mitotic figures. To overcome this imbalance and to avoid over-fitting, we performed augmentation and oversampling of patches containing mitotic figures. For this, 512×512 pixel contextual patches around the annotated mitotic/imposter figures were extracted after applying random shifts. The above sampling was repeated to create an equal number of contextual patches around both mitotic figures and imposters. Standard augmentation methods such as horizontal flipping, vertical flipping, random rescaling, random cropping, and random rotation were additionally performed to make the model invariant to geometric perturbations across different tissue types. Furthermore, Random HSV [8] transformation was also applied to randomly change the hue, saturation, and value (HSV) representation of images, for making the model robust to color perturbations caused by variability in image acquisition and protocols. Further, following the same approach, an equal number of mitotic and imposter contextual patches of size 96×96 pixels were also separately generated.

2.3   Ensemble Model

Our ensemble model has two components, namely, a visual transformer-based object detector and a CNN based classifier.

Mitosis Detection Model (MD): We use Detection Transformer (DETR) [9] as the detection model, trained with the 512×512 contextual patches. Resnet50-DC5 *backbone* pretrained on ImageNet was selected for DETR based on ablation experiments involving various popular backbones. The AdamW optimizer [10] was used as the optimization method for model training. The initial learning rate was set to 1e–04 and reduced after every 30 epochs. The batch size was chosen as 6. Cross Entropy (CE) loss function was used for the loss, as the dataset is balanced between mitotic and imposter classes. The overall loss was composed of boundary loss and CE loss. The hyper-parameters of DETR were chosen using Optuna hyper-parameter optimization framework [11]. In the final submission, the model was trained on both the training and validation data for 65 epochs.

Mitosis Classification Model (MC): EfficientNet-B7 [12] pretrained on ImageNet was used as the classifier with BCE loss. Adam optimizer [13] was used with an initial learning rate of 1e–03. The model was trained on the dataset of 96×96 mitotic and imposter contextual patches (see Data pre-processing) with an 80:20 split. The hyper-parameters of the EfficientNet-B7 were chosen using Optuna hyper-parameter optimization framework. In the final submission, the model was trained on both the training and validation data for 35 epochs.

Ensemble Model: For the test WSI images, using a sliding window with an overlay of 50 pixels, cropped patches of size 512×512 pixels were extracted. Each cropped patch was first fed to the trained DETR model. The DETR output consists of the identified objects and their associated class probabilities. All mitotic object predictions with class probability ≥ 0.9 were directly included in the final output. This threshold was chosen based on the performance on the validation data. For the remaining mitotic predictions (with classification probability <0.9), we extracted smaller patches of 96×96 pixels around each such prediction within the 512×512 patch and input them to the classification model. The final class probabilities for these objects were obtained by averaging the DETR and classifier prediction probabilities. After experimenting with different threshold values, a threshold of 0.75 was chosen and mitotic figures whose aggregate prediction probabilities exceeded 0.75 were additionally included in the final output.

Our models were implemented using Pytorch framework and all experiments were run on Nvidia DGX A100.

3   Results and Discussion

Table 1 shows the comparison of F1-scores of our approach on the validation, preliminary (4 tumor types) and final (10 tumor types) datasets. The results are for Task 1 of the MIDOG 2022 Challenge [6]. The MD row shows the performance when our mitosis detection model alone was used. Our ensemble model secured second position in the challenge [14].

Table 1. F1 scores on the challenge data - Task 1

Model	F1 (Validation Set)	F1 (Preliminary Set)	F1 (Final Set)
MD model	0.7906	0.7607	0.7510
Ensemble (Our model)	0.8093	**0.7704**	**0.7569**
Baseline 2	-	0.7152	0.7135
Baseline 1	-	0.6285	0.6542

3.1   Ablation Studies

We conducted extensive ablation studies to understand the effect of different parameters on the final model performance. This included experiments on different patch sizes (512 vs 128 pixels), stain augmentation vs stain normalization, choice of DETR backbones (ResNet50, ResNet50-DC5, ResNet101 and ResNet101-DC5), Classifier CNN models (ResNet, Densenet, EfficientNet-B7, EfficientNet-B8. EfficientNet-B7) and deep supervision where Hematoxylin channel was added as a 4th channel to the images. Based on these experiments, we chose a patch size of 512 pixels, Random HSV for stain augmentation, ResNet50-DC5 as DETR CNN backbone and EfficientNet-B7 as the classifier. However, adding a Hematoxylin channel did not improve the performance on the validation set. All experiments are repeated under the same training and evaluation conditions.

Inclusion of Unlabeled Tumor: We also included the patches from the unlabeled tumor tissue for training the classifier. For this, the high confidence MD model predictions were considered as pseudo labels for training. Training dataset was created using the four labeled tumor types and the unlabeled tumor type (human melanoma). The validation dataset was created from the human neuroendocrine tumor tissue. However, no significant improvement in the performance was observed on the validation set.

Our work shows that visual transformers show promise in solving complex imaging problems arising in digital pathology.

References

1. Dawson, H.: Digital pathology - Rising to the challenge. Front. Med. (Lausanne) **9** 888896, (2022)
2. Jiménez, G., Racoceanu, D.: Classification in computational pathology: application to mitosis analysis in breast cancer grading. Front Bioeng Biotechnol **7**(145), (2019)
3. Veta, M., et al.: Predicting breast tumor proliferation from whole-slide images: the TUPAC16 challenge. Med. Image Anal. **54**(145), 111–121 (2019)
4. Mitos & Atypia 14 contest home page, https://mitos-atypia14.grand-challenge.org/home/. Accessed 4 Sept 2022
5. Aubreville, M., et al.: Mitosis domain generalization in histopathology images - the MIDOG challenge. Med. Image Anal. **84**, 102699 (2023)
6. Aubreville, M., et al.: Mitosis Domain generalization challenge. In: 2022–25th International Conference on Medical Image Computing and Computer Assisted Intervention (MICCAI 2022). Zenodo. https://doi.org/10.5281/zenodo.6362337
7. Aubreville, M., et al.: MItosis DOmain generalization challenge 2022 (MICCAI MIDOG 2022), Training data set (PNG version) (1.0) [Data set]. Zenodo. https://doi.org/10.5281/zenodo.6547151
8. Tellez, D., et al.: Quantifying the effects of data augmentation and stain color normalization in convolutional neural networks for computational pathology. Med Image Anal. **58**, 101544 (2022). Epub 2019 Aug 21. PMID: 31466046. https://doi.org/10.1016/j.media.2019.101544
9. Carion, N., et al.: End-to-End Object Detection with Transformers. arxiv (2005.12872v3), (2020)
10. Adam, W.: https://arxiv.org/abs/1711.05101. Accessed 29 Aug 2022
11. Akiba, T., Sano, S., Yanase, T., Ohta, T., Koyama, M.: Optuna: A next-generation hyperparameter optimization framework. In: Proceedings: 25th ACM SIGKDD International Conference on Knowledge Discovery and Data Mining (2019)
12. Tan, M., Le, QV.: EfficientNet: rethinking model scaling for convolutional neural networks. In: Proceedings: ICML (2019)
13. Kingma, D., Ba, J.: Adam: a Method for stochastic optimization. In: 3rd International Conference on Learning Representations (ICLR 2015) Proceedings. ICLR, San Diego, CA, USA (2015)
14. MIDOG 2022 results. https://midog2022.grand-challenge.org/evaluation/final-test-phase-task-1-without-additional-data/leaderboard/. Accessed 21 Sept 2022

Fine-Grained Hard-Negative Mining: Generalizing Mitosis Detection with a Fifth of the MIDOG 2022 Dataset

Maxime W. Lafarge[✉] and Viktor H. Koelzer

Department of Pathology and Molecular Pathology,
University Hospital and University of Zurich, Zurich, Switzerland
Maxime.Lafarge@usz.ch

Abstract. Making histopathology image classifiers robust to a wide range of real-world variability is a challenging task. Here, we describe a candidate deep learning solution for the Mitosis Domain Generalization Challenge 2022 (MIDOG) to address the problem of generalization for mitosis detection in images of hematoxylin-eosin-stained histology slides under high variability (scanner, tissue type and species variability). Our approach consists in training a rotation-invariant deep learning model using aggressive data augmentation with a training set enriched with hard negative examples and automatically selected negative examples from the unlabeled part of the challenge dataset. To optimize the performance of our models, we investigated a hard negative mining regime search procedure that lead us to train our best model using a subset of image patches representing 19.6% of our training partition of the challenge dataset. Our candidate model ensemble achieved a F_1-score of .697 on the final test set after automated evaluation on the challenge platform, achieving the third best overall score in the MIDOG 2022 Challenge.

Keywords: mtosis detection · domain generalization · MIDOG 2022

1 Introduction

To support the research community with the development of new mitosis detection algorithms that are robust to scanner variability, the *MIDOG 2021 Challenge* [2] lead to an overview of efficient approaches towards solving this task. To further encourage the development of models that can generalize beyond inter-scanner variability, the *MIDOG 2022 Challenge* was initiated [1], offering a unique opportunity to compare the generalization ability of mitosis detectors in a blind manner, as the challenge organizers independently evaluate candidate solutions on held-out sets of images from undisclosed and unseen domains.

This opportunity motivated us to revisit the pioneer methodology proposed by Cireşan et al. [3], and to assess the relative performance of standard methods in the context of modern deep convolutional neural network architectures and large high-variability mitosis datasets as the one provided for this challenge.

© The Author(s), under exclusive license to Springer Nature Switzerland AG 2023
B. Sheng and M. Aubreville (Eds.): MIDOG 2022/DRAC 2022, LNCS 13597, pp. 226–233, 2023.
https://doi.org/10.1007/978-3-031-33658-4_24

Our approach consists of a patch-based training procedure that we used to train deep learning models that can then be applied to detect mitoses on unseen images, building upon the strategy we employed for the *MIDOG 2021 challenge* [8]. We incrementally made changes to this training procedure over the development phase of the challenge and monitored performances on our validation partition of the challenge dataset in order to select and submit our best performing model. In this paper, we describe the different components that constitute our submitted solution, including a fine-grained assessment of a hard negative mining procedure that we consider to be the main contributing part of our solution.

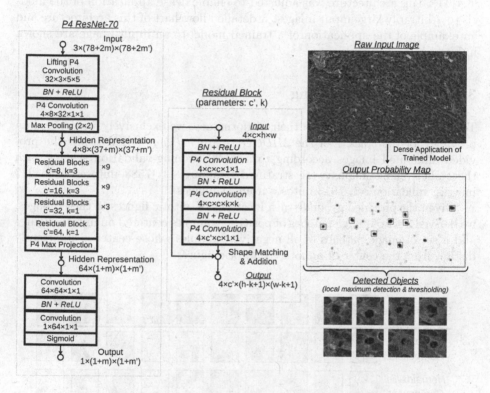

Fig. 1. Architecture of the 70-layer ResNet used in this work and example of its application to a raw input H&E histopathology image. The shape of output tensors is written with the following format: (*Orientations×*)*Channels×Height×Width*. The shape of trainable operator tensors is written with the following format: (*Orientations×*)*Out.Channels×In.Channels×Ker.Height×Ker.Width*.

2   Model Architecture

We implemented a customized 70-layer ResNet architecture [6] to model the confidence probability for input images to be centered on a mitotic figure within a receptive field of 78×78 pixels. We replaced standard convolutional layers by P4-group convolutional layers [4] to guarantee invariance of our models to 90-degree rotations without requiring train-time or test-time rotation augmentations, with a low computational overhead. This change was further motivated by the improvment of performance across multiple histopathology image classification tasks for models using this type of operation reported in the literature [5, 7, 11]. This architecture was adjusted to enable dense application of the models to arbitrarily large input images. A detailed flowchart of this architecture and an example of the application of a trained model to an input image are shown in Fig. 1.

3   Dataset Partitioning

To train models and evaluate their performance, we exclusively used the data provided for the track 1 of the *MIDOG 2022 Challenge* [1]. We split the provided annotated images according to a 80-20 training-validation scheme such that labels and domains were stratified (training set: 7588 mitoses from 283 images; validation set: 1913 mitoses from 71 images).

Given the provided ground-truth locations of mitotic figures in these images, we derived a set of image patches of positive examples centered on mitotic figures and a set of image patches of all negative examples whose center is sufficiently distant from the center of annotated mitotic figures.

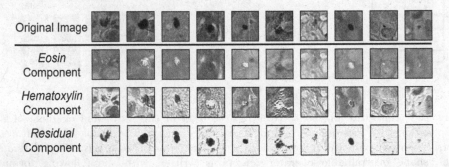

Fig. 2. Example of automatically selected image patches from the unlabeled part of the *MIDOG 2022* dataset based on the optical density of their residual component after application of a stain unmixing procedure. These selected image patches were used to enrich the training set.

4   Training Procedure

All models were trained by minimizing the cross-entropy loss via stochastic gradient descent with momentum (initial learning rate 0.03 and momentum 0.9) using input mini-batches of size 128. We used a cyclic learning rate scheduling [9] with a cycle period of 10k iterations and applied weight decay regularization (coefficient 10^{-4}). Mini-batches were generated with randomly sampled image patches of size 78×78, balanced between positive and negative examples. All image patches were randomly transformed according to an augmentation protocol (including channel-wise intensity distortions) whose operations and parameter ranges are detailed in Table. 1. For evaluation purposes, we saved the weights of the model that achieved the lowest validation loss within 150k training iterations.

Table 1. Data augmentation protocol: for each input image patch, the following list of transformations is scanned and applied with a given probability. Transformation parameters are randomly sampled in a given interval. The two variants of the protocol used to train our submitted model ensemble are detailed here.

Transformation	Policy A		Policy B	
	Coefficients	Probability	Coefficients	Probability
Transposition	–	50%	–	50%
Elastic Deformation	–	100%	–	50%
Spatial Shift ($\Delta_{x,y}$)	[−12px, 12px]	100%	[−12px, 12px]	100%
Spatial Zoom (α)	[−10%, 20%]	50%	[−10%, 20%]	50%
(HLS) Hue Rotation (h)	[0°, 360°]	80%	[−60°, 60°]	50%
Color Shift ($c_{r,g,b}$)	[−51, 51]	80%	[−151, 51]	50%
Contrast Correction ($\mu_{r,g,b}$)	[0.8, 1.2]	80%	[0.8, 1.2]	50%
Gamma Correction ($\gamma_{r,g,b}$)	–	0%	[0.8, 1.2]	50%

5   Fine-Grained Hard Negative Mining

Hard negative mining (HNM) has become a standard procedure for the development of mitosis detectors since the solution proposed by [3], and aims at improving the model performance by using a well-chosen subset of "hard" negative examples for training instead of using randomly sampled negative examples. This procedure typically requires training a first model with all the available positive and negative examples, then ranking all negative examples based on the confidence score output by this trained model, and finally keeping the negative examples with a score above a fixed cutoff threshold as a set of "hard negatives" to be used to repeat training and improve performances.

For this challenge, we considered this threshold as a hyper-parameter and searched for an optimal number of hard negative examples to select that would maximize performance on the validation set. A summary of the binary search we conducted for this parameter is shown in Fig. 3.

This fine-grained assessment enabled the selection of an optimal subset of 121 − 738 hard negative examples of our training partition, which improved validation performances in comparison to using randomly sampled training examples. Our submitted model ensemble was trained using this optimal subset of image patches along with all positive examples whose joint total pixel count represents 19.6% of the overall pixel count of our training partition.

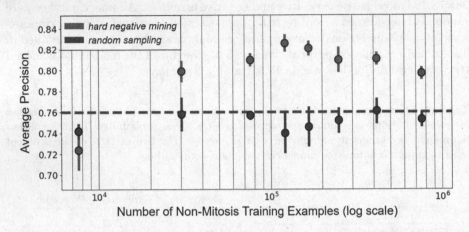

Fig. 3. Average precision (AP) on the validation set of models trained with subsets of non-mitosis examples of different sizes. These subsets were either generated via random sampling or by hard negative mining. During training, the positive class is oversampled to ensure balance of classes in training batches. Circles and bars represent the mean and standard deviation of AP for three repeated experiments with different random seeds. The dashed line indicates the mean AP obtained when training using all the possible negative examples of our training partition.

6   Automated Enrichment with Negative Examples

To further enrich the dataset with additional negative examples we implemented a stain unmixing algorithm derived from [10] to first separate the hematoxylin, eosin and residual components in the unlabeled images of the dataset, and then automatically extract image patches with high optical density in the estimated residual component to enrich the training set with extra negative examples. This procedure resulted in the selection of 72 negative examples after application of an optimal hard negative mining selection rule as described in the previous section. A batch of such selected examples is shown in Fig. 2. This approach was motivated by the observation that true positive mitotic events reside in defined stain vectors and are separable from background events such as pigmentation and ink that form common impostors.

7   Inference Pipeline

Once our models were trained, we produced prediction maps by applying them densely on test images, and then derived candidate detection locations as the set of local maxima. We then considered all candidate locations with a prediction score above a threshold value as final detection locations (this threshold was chosen as the one maximizing the F_1-score on the validation set).

For our final submission, we created a model ensemble by taking the agreement between the detection sets of two models trained under two variants of our augmentation protocol as detailed in Table 1. A comparison of performance of our models against baselines is summarized in Table 2.

Table 2. Comparison of F_1-scores of our trained models (two different augmentation policies and their ensemble) and baselines of the *MIDOG 2022 Challenge* on the four tumor types (*T1,2,3,4*) of the hidden preliminary test set and on the final test set of the challenge.

Model	Internal		Preliminary Test				Final Test
	Validation	Overall	T1	T2	T3	T4	Overall
ours (*Policy A*)	.784	.646	**.783**	.726	.548	.708	–
ours (*Policy B*)	.787	.571	.757	**.775**	.428	.571	–
ours (Ensemble)	**.791**	.690	.758	.735	.653	.610	.696
MIDOG 2022 (Baseline 2)	–	**.715**	.744	.732	**.692**	.711	**.714**
MIDOG 2022 (Baseline 1)	–	.629	.753	.608	.585	**.743**	–

8   Discussion

We present a candidate algorithm that achieves moderate generalization performance with an overall F_1-score of .696 on the final test set of the challenge (set of images from ten unseen domains), while relying mostly on conventional methods (patch-based training procedure, ResNet architecture, cross-entropy loss minimization, hard negative mining, standard augmentation protocol).

Performance on the different domains of the test sets were heterogeneous indicating that the investigated pipeline only enables generalization to a subset of the held-out test domains of the challenge.

Hard Negative Mining played an important role in achieving the final performance of our submitted solution: the search for an optimal amount of hard negative training examples helped improving performances on the validation set with an increase of Average Precision from .760 to .826. Our comparative analysis suggests that this level of performance could not have been achieved from training using only arbitrary negative examples.

Furthermore, we investigated the use of a conventional stain unmixing method to automatically extract potential negative examples in unlabeled images. As this approach did not decrease the performances on the validation set, we assumed this helped our models better generalizing. We thus used the resulting enriched training set to train our final models, yet, because of the very small number of training examples selected this way, we cannot further conclude whether the effect of this strategy was beneficial or not regarding the performance on the final test set. Because this enrichment procedure relies on the strong assumption that extracted examples belong to the non-mitosis class, it should be kept in mind that positive examples could still potentially be extracted. Here, we recommend that the examples extracted by such a method should ideally be reviewed by expert annotators to ensure the correctness of the class of these additional training examples.

Although the reported variations of our augmentation protocol produced similar results on the validation set, we observed that they produced heterogeneous performances on the preliminary test set (Table 2). This suggests that generalization to the domains of the preliminary test set is sensitive to variations of the augmentation protocol used for training. Indeed, the two reported augmentation policies (Table 1) helped generalizing to different parts of the unseen variability of the preliminary test set (*Policy A* enabled better generalization for Tumor 1 whereas *Policy B* enabled better generalization for Tumor 2). These results corroborate the known association between chosen augmentation policies and generalization performance: this motivates us to further study how to best design and select augmentation protocols to improve domain generalization.

References

1. Aubreville, M., Bertram, C., Breininger, K., Jabari, S., Stathonikos, N., Veta, M.: MItosis DOmain Generalization challenge 2022. In: Proceedings of the International Conference on Medical Image Computing and Computer-Assisted Intervention (MICCAI) (2022). https://doi.org/10.5281/zenodo.6362337
2. Aubreville, M., et al.: Mitosis domain generalization in histopathology images - the MIDOG challenge. Med. Image Anal. **84**, 102699 (2023)
3. Cireşan, D.C., Giusti, A., Gambardella, L.M., Schmidhuber, J.: Mitosis detection in breast cancer histology images with deep neural networks. In: Proceedings of the International Conference on Medical Image Computing and Computer-Assisted Intervention (MICCAI) (2013)
4. Cohen, T., Welling, M.: Group equivariant convolutional networks. In: Proceedings of the International Conference on Machine Learning (ICML). pp. 2990–2999 (2016)
5. Graham, S., Epstein, D., Rajpoot, N.: Dense steerable filter CNNs for exploiting rotational symmetry in histology images. IEEE Trans. Med. Imaging **39**, 4124–4136 (2020)
6. He, K., Zhang, X., Ren, S., Sun, J.: Identity mappings in deep residual networks. In: Leibe, B., Matas, J., Sebe, N., Welling, M. (eds.) ECCV 2016. LNCS, vol. 9908, pp. 630–645. Springer, Cham (2016). https://doi.org/10.1007/978-3-319-46493-0_38

7. Lafarge, M.W., Bekkers, E.J., Pluim, J.P., Duits, R., Veta, M.: Roto-translation equivariant convolutional networks: application to histopathology image analysis. Med. Image Anal. **68**, 101849 (2021)

8. Lafarge, M.W., Koelzer, V.H.: Rotation invariance and extensive data augmentation: A strategy for the MItosis DOmain Generalization (MIDOG) challenge. In: International Conference on Medical Image Computing and Computer-Assisted Intervention (2021)

9. Loshchilov, I., Hutter, F.: SGDR: Stochastic gradient descent with warm restarts. In: International Conference on Learning Representations (ICLR) (2017)

10. Macenko, M., et al.: A method for normalizing histology slides for quantitative analysis. In: Proceedings of the IEEE International Symposium on Biomedical Imaging (ISBI) (2009)

11. Macenko, M., et al.: A method for normalizing histology slides for quantitative analysis. In: Proceedings of the IEEE International Symposium on Biomedical Imaging (ISBI) (2009)

Multi-task RetinaNet for Mitosis Detection

Ziyue Wang[2(✉)], Yang Chen[1], Zijie Fang[2], Hao Bian[1], and Yongbing Zhang[1]

[1] Department of Computer Science, Harbin Institute of Technology, Harbin, China
[2] Shenzhen International Graduate School, Tsinghua University, Beijing, China
`200111326@stu.hit.edu.cn`

Abstract. The count of mitotic cells is a key feature in tumor diagnosis. However, due to the variability of mitotic cell morphology, detecting mitotic cells in tumor tissues is a highly challenging task. At the same time, the performance of the trained models often declines when there is a vast difference between the source domain and the target domain. (i.e., the different tumor types and scanners). Therefore, it is necessary to develop algorithms for detecting mitotic cells with robustness in domain shift scenarios. Our work proposes a foreground detection and tumor classification task based on the baseline (Retinanet) and utilizes data augmentation to improve our model's detection ability and domain generalization performance. We achieve excellent performance on the challenging preliminary test dataset (F1 score: 0.5809) and the Final test dataset (F1:0.6300).

Keywords: Object Detection · Mitosis Detection · Multi tasks

1 Introduction

Histopathology is the gold standard for tumor diagnosis and prognosis. With the rapid development of deep learning technology and the popularization of whole slide image (WSI) scanners, computational pathology has received significant attention in recent years.

In the clinical diagnosis of tumors, the account of mitotic cells is an essential feature for judging the degree of malignant proliferation of tumors, which is a critical prognostic indicator and reference for tumor grading and treatment.

In recent years, many methods have succeeded in mitotic cell detection with the development of deep learning in computer vision. However, due to the variability of mitotic cell morphology, it has become an increasing challenge to identify mitotic cells directly from H&E-stained histopathological sections, especially when there is a domain gap between the training data and testing data (i.e., the data is from different tumor types or different scanners). Therefore, new deep learning methods with robustness in domain shift scenarios are required.

Z. Wang and Y. Chen—Contributed equally.

B. Sheng and M. Aubreville (Eds.): MIDOG 2022/DRAC 2022, LNCS 13597, pp. 234–240, 2023.
https://doi.org/10.1007/978-3-031-33658-4_25

2   Material and Methods

2.1   Dataset Description

Mitosis domain generalization challenge 2022 (MIDOG 2022) [1,2] is a further expansion of tumor types, and types of scanners based on the MIDOG 2021 competition [3–5]. In the MIDOG 2021 competition, the competition only focused on the color generalization error caused by different scanners on the same tumor (breast cancer). In MIDOG 2022, the competition organizers digitized different tumor cell sections from different laboratories and species (i.e., human, canine) under different scanners. MIDOG 2022 aims to explore mitosis cell detection algorithms with multiple domains (differences in tissue and collection methods).

The training dataset contains 405 WSIs, including six tumor types and two species: Canine Lung Cancer, Human Breast Cancer, Canine Lymphoma, Human Neuroendocrine Tumor, Canine Cutaneous Mast Cell Tumor, and Human Melanoma. The WSIs are scanned with three different scanners: 3DHistech Pannoramic Scan II, Hamamatsu NanoZoomer XR, and Aperio ScanScope CS2. Organizers of the challenge set a preliminary test for the self-evaluation of participating pipelines, which contain 20 WSIs from two tumor types and two species: Human Breast Cancer, Human Lymphoma, Canine Osteosarcoma, and Canine Pheochromocytoma.

The test dataset contains 100 WSIs, contains ten kinds of tumor types, and has three species, including Human Melanoma, Human Astrocytoma, Human bladder Carcinoma, Canine Breast Cancer, Canine cutaneous Mast Cell Tumor, Human Meningioma, Human Colon Carcinoma, Canine Hemangiosarcoma, Feline Soft Tissue Sarcoma, Feline Lymphoma. WSIs from unseen tumor types, scanners, labs, and species can well reveal the capability of participating algorithms to generalize to various tumors.

The mitotic cells of the competition are determined by visual assessment by a trained pathologist. The competition organizer provided mitotic cell annotations for the five tumor types in the training dataset, excluding Human melanoma, and the corresponding scanner type for each tumor type. Since breast cancer contains all three scanner types simultaneously, and no specific scanner type corresponds to each WSI, we consider that the scanner type of the breast cancer tissue section is unknown.

2.2   Methods

We propose a multi-task mitotic cell detection model based on RetinaNet, in which the main improvements include three parts:

– An auxiliary classification network is used to classify tumors, including six categories: prostate cancer, lymphoma, lung cancer, melanoma, breast cancer, and mast cell tumor.
– An auxiliary classification network is used to classify whether the patch contains mitotic cells (or hard samples).

– A data augmentation transform is used to improve the domain generalization ability of the model and detect mitotic cells of different types of tumors (different scanners).

2.3   The Choice of Baseline

To select the best baseline for mitotic cell detection, we compare a set of state-of-the-art object detection algorithms and choose RetinaNet as our baseline for its excellent performance in the challenge.

2.4   Multi Task Auxiliary Classification

As shown in Fig. 1, we first select the deep feature of FPN in RetinanNet [6] and then add two fully connected network auxiliary classification tasks after this deep feature.

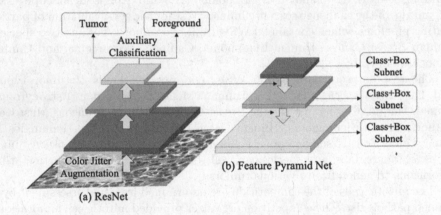

Fig. 1. Overview of multi-task RetinaNet.

For the tumor auxiliary classification task, we use cross entropy. During the training, we consider 6 tissue types at the same time. The loss function is defined as follows:

$$CE\left(p_c\right) = -\sum_{i=0}^{C-1} y_i \log\left(p_i\right) = -\log\left(p_c\right), \qquad (1)$$

where, y_i represents the real tumor category of the ith object, p_i indicates the predicted tumor category probability, and C indicates the number of tumor categories. By adding the auxiliary task, the encoder can learn to encode different tumor types, thus improving the detection ability. Even if the tumor types in the test sets are unseen, some tumor types are similar in morphology to those in the training set (i.e., the canine lymphoma in the training set and the feline lymphoma in the final test set). The model extracts features similar to certain learned tumor types in the training set for unknown types in the test set, and the detection ability can keep excellent. The effect of improving the basic detection ability of the model overweights the effect of losing domain generalization, thus improving the overall performance.

For the foreground auxiliary classification task, we firstly regard the patches containing mitotic cells or hard samples as foreground classes, and the patches that do not contain mitotic cells or hard samples as background classes, and use Focal loss [6] as the loss function, which is defined as follows:

$$FL\left(p_c\right) = -\alpha_c \left(1 - p_c\right)^{\gamma} \log\left(p_c\right), \qquad (2)$$

where α and γ are two super parameters of focal loss, this task can help the model identify whether there are mitotic cells (hard samples) in the patches. The features are unchanged when the domain shifts. Adding the auxiliary task can improve the situation of false positive samples and false negative samples, thus improving the overall F1 score.

We generate the ground truth of the class labels for the two auxiliary tasks for the labeled data. In the training phase, the network's input is the patches, and the output is the bounding boxes for detection tasks and the class labels for the classification tasks. We only keep the bounding boxes as the output in the testing phase.

2.5   Scanner Differences and Data Augmentation

Adding the two auxiliary tasks improves the model's performance when the domain shifts due to different tissues or tumor types. For the domain gap caused by different scanners, we have tried two other solutions:

Firstly, we add a gradient-reverse layer after the encoder, followed by a classification head as a domain discriminator. However, in our experiments, the method shows little effect in improving the model's generalization ability. The F1 score even drops by 0.003 when testing on the target domain.

Thus, we use the Color Jitter data enhancement module to improve the domain generalization ability of the model, where the parameter settings of Color Jitter are: brightness = 0.35, contrast = 0.2, saturation = 0.1, hue = 0.1. Other data augmentation includes random crop, horizontal flip, and vertical flip. These data augmentations can help improve the robustness of the domain gap caused by different scanners. The F1 score on the target domain increased by 0.015, shown in Table 2.

2.6   Hyperparameter Setting

We conduct our model under the pytorch framework. We use Adam optimizer. The learning rate is $1e-5$, and the batch is 16. For RetinaNet's FPN, we use the ResNet 50 network based on ImageNet 1K pre-training. For the hyperparameter of Focal loss, we set $\alpha = 0.25$ and $\gamma = 2$.

3 Results

To verify the mitotic cell detection performance and generalization performance of our proposed model, we re-divided the training set and validation set according to the original training set. The newly constructed training set includes four tumor tissues: canine lung cancer, human breast cancer, canine lymphoma, and canine cutaneous mastocytoma. The test set comprises human neuroendocrine tumors. The tumor type and the scanner type of the test set are unseen; thus, the study can show our model's generalization ability when meeting unknown domains. Under the general hyperparametric setting, we compare the performance of different baselines and verify the performance of multi-task RetinaNet on the validation set through ablation studies.

3.1 Baseline Model Selection

In order to better realize the detection of mitotic cells, we first need to select the algorithm which is most suitable for the detection of the mitotic cells from a series of classical object detection algorithms. As shown in Table 1, among all the main object detection algorithms, RetinaNet shows the best performance on mAP, mAR, and F1 score (iou is 0.5). Therefore, we choose RetinaNet as the baseline model for cell detection tasks and additionally add multi-task learning branches on RetinaNet to further improve the domain generalization ability.

Table 1. The performance of baseline models in the detection of the mitotic cells

model	mAP (iou = 0.5)	mAR (iou = 0.5)	F1(iou = 0.5)
Yolo_v3 [7]	0.395	0.558	0.463
Faster R-CNN [8]	0.369	0.566	0.447
Mask R-CNN [9]	0.422	0.533	0.471
CornerNet [10]	0.361	0.396	0.378
CenterNet [11]	0.366	0.374	0.370
RetinaNet [6]	0.433	0.615	0.508

3.2 Ablation Study

We split the training set into two domains to conduct the ablation study. We use WSIs from Canine Cutaneous Mast Cell Tumors scanned with Aperio ScanScope CS2 as the target domain and the rest as the source domain. We train our model on the source domain and validate it on the target domain. Both the tumor type and the scanner type of the target domain are unseen; thus, the study can show our model's generalization ability when meeting unknown domains.

As shown in Table 2, the performance of RetinaNet's F1 score (iou is 0.5) on mitosis cell detection is 0.508. Compared with the baseline, the three components proposed above can improve the detection performance significantly. The most noticeable performance improvement comes from the foreground auxiliary classification component. When we integrate all three proposed components, the model's performance reaches a 0.568 F1 score on the validation set. Interestingly, the F1 score is 0.581 when tested on the primary dataset, which shows the excellent generalization ability of our method.

3.3   Result on the Final Test Set

The overall F1 score on the final test is 0.6300, the mAP is 0.4317, the recall is 0.5345, and the precision is 0.7670 (iou is 0.5), which means the false negative samples affect the model performance more than the false positive samples. Our model works well on most tumor types. However, on tumor type 1, the F1 score is only 0.3750, and the precision is only 0.2523. Furthermore, the F1 score (0.5548, 0.5676, 0.4910 separately) and the recall(0.4272, 0.5676, 0.3534 separately) of tumor type 3, 5, 8 is also low. The generalization ability of our model still needs to be improved to perform well on specific tumor types.

Table 2. Ablation studies on multi-task RetinaNet.

foreground classification	tumor classification	data augmentation	mAP (iou = 0.5)	mAR (iou = 0.5)	F1 (iou = 0.5)
			0.433	0.615	0.508
		✓	0.449	0.626	0.523
	✓		0.458	0.638	0.533
✓			0.451	0.657	0.534
	✓	✓	0.463	0.642	0.538
✓		✓	0.465	0.669	0.549
✓	✓		0.473	0.672	0.555
✓	✓	✓	0.486	0.682	0.568

4   Discussion

Our research shows that for the mitotic cell detection task, adding auxiliary classification loss helps to improve the detection performance of the model significantly when the data are from various tumor types. For the domain generalization problem caused by different color disturbances of scanners, using Color Jitter data augmentation can also increase the domain generalization performance and avoid high training costs from adversarial techniques. The model can also be robust when the test images come from unknown domains.

References

1. Aubreville, M., Bertram, C., Breininger, K., Jabari, S., Stathonikos, N., Veta. M.:MItosis DOmain generalization challenge 2022. In: 25th International Conference on Medical Image Computing and Computer Assisted Intervention (MICCAI 2022). Zenodo. https://doi.org/10.5281/zenodo.6362337
2. Aubreville, M., et al.: Mitosis domain generalization in histopathology images - The MIDOG challenge, Med. Anal **84** (2022)
3. Aubreville, M., Zimmerer, D., Heinrich, M. (eds). Biomedical image registration, domain generalisation and out-of-distribution analysis: MICCAI 2021 Challenges: MIDOG 2021, MOOD 2021, and Learn2Reg 2021, Held in Conjunction with MICCAI 2021, Strasbourg, France, September 27–October 1, 2021, Proceedings, volume 13166, LNCS. Springer, Cham (2022). ISBN 978-3-030-97280-6 978-3-030-97281-3. https://doi.org/10.1007/978-3-030-97281-3
4. Aubreville, M., et al.: MItosis DOmain generalization challenge. In: 24th International Conference on Medical Image Computing and Computer Assisted Intervention (MICCAI 2021). Zenodo. https://doi.org/10.5281/zenodo.4573978
5. bibitemch25midog2021MedIA Aubreville, M., et al.: Mitosis domain generalization in histopathology images – The MIDOG challenge, Med. Image Anal. **84**, 102699 (2023). ISSN: 1361-8415. https://doi.org/10.1016/j.media.2022.102699
6. bibitemch25midog2021MedIA Aubreville, M., et al.: Mitosis domain generalization in histopathology images – The MIDOG challenge, Med. Image Anal. **84**, 102699 (2023). ISSN: 1361-8415. https://doi.org/10.1016/j.media.2022.102699
7. Redmon, J., Farhadi, A.: Yolov3: an incremental improvement. arXiv preprint arXiv:1804.02767 (2018)
8. Girshick, R.: Fast R-CNN. In: Proceedings of the IEEE International Conference on Computer Vision (2015)
9. He, K., Gkioxari, G., Dollár, P., Girshick, R.: Mask R-CNN. In: Proceedings of the IEEE International Conference on Computer Vision (2017)
10. Law, H., Deng, J.: CornerNet: detecting objects as paired keypoints. In: Ferrari, V., Hebert, M., Sminchisescu, C., Weiss, Y. (eds.) Computer Vision – ECCV 2018. LNCS, vol. 11218, pp. 765–781. Springer, Cham (2018). https://doi.org/10.1007/978-3-030-01264-9_45
11. Duan, Kaiwen and Bai, Song and Xie, Lingxi and Qi, Honggang and Huang, Qingming and Tian, Qi. Centernet: Keypoint triplets for object detection. Proceedings of the IEEE/CVF international conference on computer vision. 2019

Author Index

© The Editor(s) (if applicable) and The Author(s), under exclusive license
to Springer Nature Switzerland AG 2023
B. Sheng and M. Aubreville (Eds.): MIDOG 2022/DRAC 2022, LNCS 13597, pp. 241–242, 2023.
https://doi.org/10.1007/978-3-031-33658-4

Printed in the United States
by Baker & Taylor Publisher Services